WATER WORKS 1991

A SURVEY OF GREAT LAKES/ST. LAWRENCE RIVER WATERFRONT DEVELOPMENT

How Great Lakes/St. Lawrence River cities keep their waterfronts thriving in the 1990s

by DANIEL K. RAY

The Center for the Great Lakes

© 1991 The Center for the Great Lakes

The Center for the Great Lakes views the Great Lakes/St. Lawrence River system as an irreplaceable asset which forms the foundation of a strong regional economy and quality of life. That potential can be realized only through a vigorous commitment to protecting and carefully developing this unique natural resource.

The Center for the Great Lakes is an Illinois 501 C-3 not-for-profit organization and a Canadian registered charitable organization.

Chairman: Anthony Earl
President: William Brah

Principal Author/Project Director: Daniel K. Ray
Editor/Production Manager: Paul Botts

This project was funded in part by the National Coastal Resources Research and Development Institute, Newport, Oregon (under Contracts No. 2-5618-25 and 2-5618-2), the Ameritech Foundation, and Warzyn Engineering.

ISBN #0921578-06-7

Published by Harbor House Publishers, Inc., 221 Water Street, Boyne City, Michigan 49712

Contents

The Harbour Development at Sandusky, Ohio.
McKAY DESIGN GROUP

Acknowledgements	5
About the Project	5
Executive Summary	7
Great Lakes and St. Lawrence River Waterfronts—An Overview	9
Six Case Studies	13
Keys to Successful Waterfront Development	21
City Profiles: Lake Superior	30
Lake Michigan	33
Lake Huron	45
Lake Erie	50
Lake Ontario	57
St. Lawrence River	64
Great Lakes Waterfronts at a Glance	67
Further Information on Waterfront Development	69
Notes	70
Geographic Index	71
Waterfront Development Fact Sheets	72

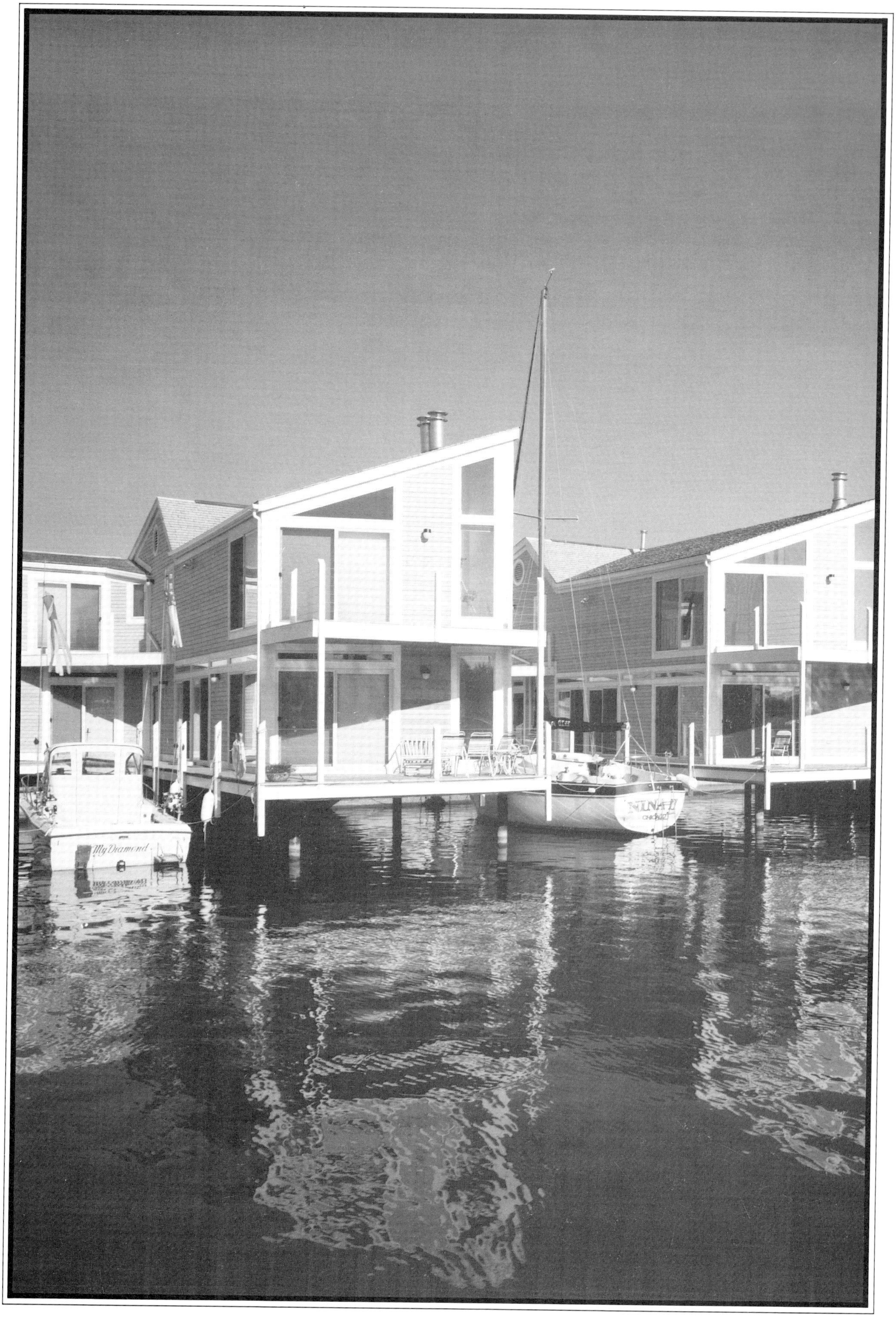
My Diamond

Acknowledgements

The Center for the Great Lakes extends its appreciation to the National Coastal Resources Research and Development Institute, the Ameritech Foundation, and Warzyn Engineering for their support of The Center's waterfront development project. The Center also recognizes the support and assistance of the International Great Lakes-St. Lawrence River Mayors' Conference, our collaborators in hosting the Water Works! 1990 conference. The mayors, and the many other speakers at the conference, provided much of the information and advice we are reporting here.

Project steering committee members provided a valuable service in overseeing and guiding the waterfront development survey work and organizing the conference. The committee included: Ann Berger, National Oceanic and Atmospheric Administration-Office of Ocean and Coastal Resource Management; Gerald Kimball, Director of Physical Planning, Duluth; Steve Thorp, Great Lakes Commission; Edward King, Director-Government and Corporate Relations, Walgreen Company; Ted Lauf, Warzyn Engineering; Richard Tobe, Commissioner, Erie County (New York) Department of Environment and Planning; Leonce Naud, Secretariat a la mise en valeur du Saint-Laurent; Ronald L. Doering, Executive Director and Counsel, Royal Commission on the Future of the Toronto Waterfront; Genevieve Ray, former executive director of the Flats Oxbow Association (Cleveland); Liz Hollander, Chicago Community Trust; and David Moore, Office of Intergovernmental Affairs, City of Milwaukee.

The Center also thanks Dave Sanders, the Center for Urban Economic Development at the University of Illinois at Chicago, and its researchers Martin Jaffe, Scott Peters, Michelle Gregory, and Adam Burck, for their assistance in the survey of waterfront developments.

Finally, The Center thanks the many officials from waterfront communities throughout the Great Lakes and St. Lawrence River region for their thoughtful contributions to the waterfront development survey and conference organization, as well as their interest and support of The Center and this project.

About the Project

The South Cove "dockominiums" in New Buffalo allow boat owners to step from their living rooms onto their decks, a concept adopted by a number of Lakes communities.
HARRY WEESE & ASSOCIATES

This report marks The Center for the Great Lakes' second study of waterfront development on the Great Lakes and St. Lawrence River. It summarizes waterfront development trends in the region and highlights some waterfront development lessons learned over the past few years. Profiles of development describe shoreline projects in 81 Great Lakes and St. Lawrence River cities. Names of key contacts and references are offered for further information. Much of the information presented is drawn from a survey of waterfront developments and from other research conducted by The Center. Other material is drawn from speakers at Water Works! 1990, a waterfront development conference held in Milwaukee May 16-18, as well as insights offered by the 250 people, including mayors of more than 50 Great Lakes/St. Lawrence River cities, who attended the conference. Parts of the text are drawn from material previously presented in the *Great Lakes Reporter,* The Center's bi-monthly newsmagazine. The report updates the information presented in The Center's 1986 Water Works! publication.

We hope that these efforts will continue to strengthen the network for waterfront development along the shores of the Great Lakes and St. Lawrence River. The region shares common opportunities and challenges in developing the shores of these freshwater seas—a bounty of redevelopable port and manufacturing lands, a vibrant boating and sportfishing industry, a four-season shoreline with snow, ice, and fluctuating lake levels, steadily improving water quality, and the desire to sustain our maritime industries and heritage. The Center's research has shown that we all have lessons to learn and tools to lend to each other as we confront these issues.

The Center, through our Great Lakes Information Service, will continue to provide information about waterfront development, including detailed fact sheets on the projects summarized in this report and a reference and referral service that links the region's waterfront development community. See page 72 for information about how to use The Center's Great Lakes Information Service to get the waterfront development information you need.

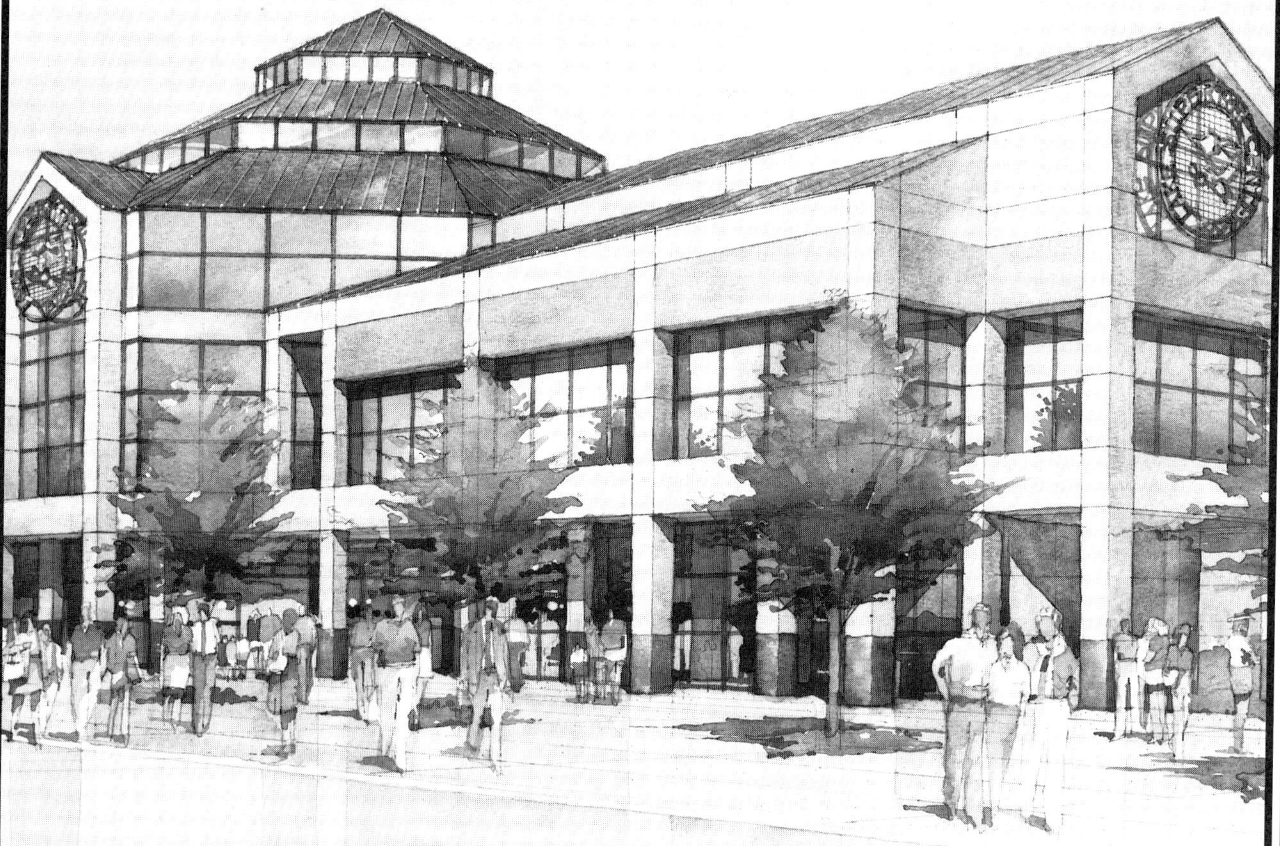

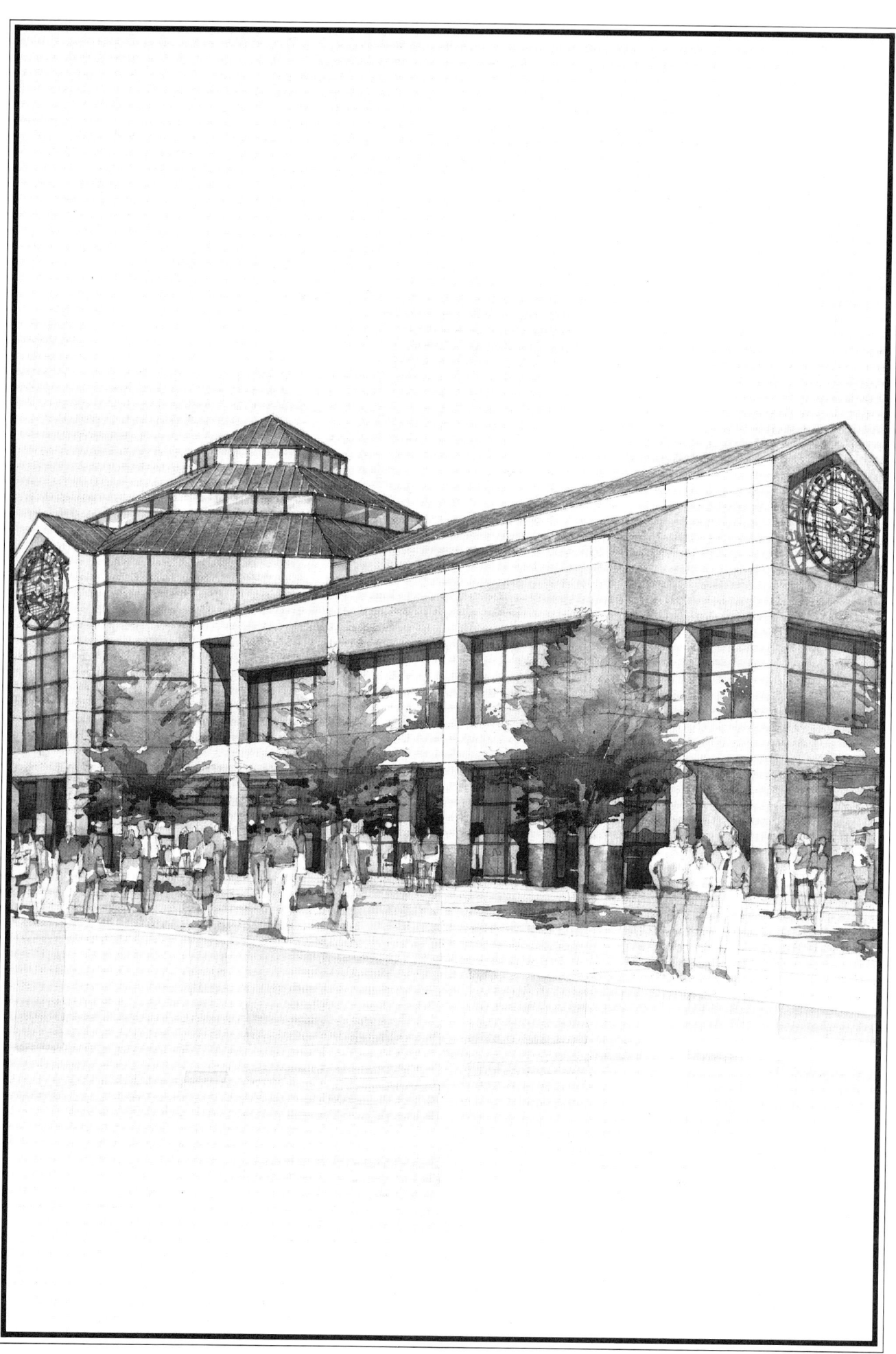

Executive Summary

A boom in waterfront development around the Great Lakes/St. Lawrence River basin in the past five years has made the region a better place to live, work and visit. Nearly 200 waterfront development projects have been planned, permitted, or constructed on the shores of 81 of the basin's largest cities since 1986. These projects have added:

- 475 acres (192 hectares) of parkland
- 37 miles (60 kilometres) of footpaths and promenades
- 13,000 boat slips
- 2,400 hotel rooms
- 7,300 new residential units
- 1.3 million square feet (387,000 square metres) of office space to the cities' waterfronts

For many cities, the waterfront development represents a reclaiming of some or all of their waterfronts from the manufacturing and transportation users which for decades occupied it. Center research found that the new development enhances property values and image of urban communities, and can provide the spark for renewal of a city's core.

Nearly 90 percent of the funding for the $3.3 billion (U.S.) of waterfront construction over the past five years has come from private sources. Public agencies, in addition to providing some funding, play key roles in assembling building sites and providing infrastructure improvements, and in some cities have taken a leading role in assuring that development includes public access to the waterfront. Provincial and state agencies have provided the bulk of the public funding for waterfront development around the basin.

Keys to successful waterfront development include promoting public access and recreation, such as including transient boat slips and boat ramps in new marinas; linking waterfront renewal and economic development; sustaining working waterfronts, and working to ensure that conflicts between traditional maritime interests and recreational users are not barriers to the use and development of the waterfront; celebrating winter, to make the waterfront a year-round amenity; building cooperation among neighboring waterfront communities; and restoring waterfront environments. The success of the region's waterfront development efforts is inextricably linked to improvements in the water quality of the Great Lakes and St. Lawrence River, and poor environmental quality can be a barrier to development and use of a waterfront.

Among the most successful waterfront revitalizations The Center found were those in which communities found unique local flavors for their waterfronts, rather than imposing "cookie-cutter" projects identical to those in other cities. *Water Works - 1991* follows up on information presented in The Center's 1986 publication *Water Works!* This new book includes surveys of development trends on each Great Lake and the St. Lawrence River, and descriptions of specific developments in each of the 81 cities surveyed.

Duluth's planned Lake Superior Center is to include a major freshwater aquarium, with exhibits focusing on the wildlife and ecology of Lake Superior and the Great Lakes (artist's rendering). LAKE SUPERIOR CENTER

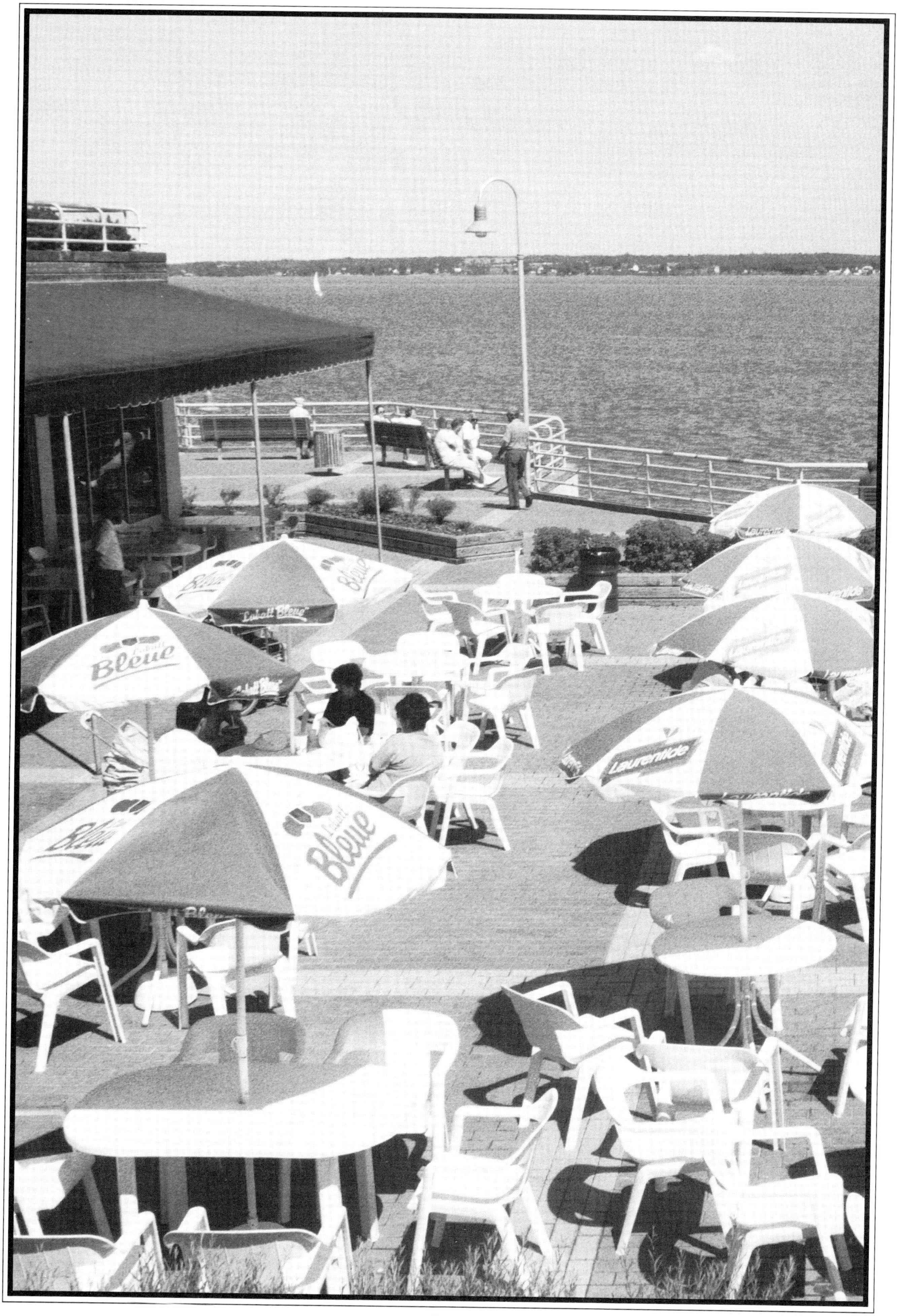
Bleue
Bleue
Laurentide

WATER WORKS

Great Lakes and St. Lawrence River Waterfronts—
An Overview

Planners in Trois-Rivieres, Quebec, reconfigured the city's waterfront to encourage strolling and sightseeing along the St. Lawrence River.
DANIEL K. RAY

As tourists stroll along a waterfront boardwalk at Duluth's Canal Park, a massive lake freighter, loaded with grain, heads out beneath the brightly lit aerial lift bridge. Inside the harbor, charter fishing boats sway at their moorings near the newly opened convention center, and music blares from a nightclub in a renovated factory building. Further along the shore local youngsters, fishing poles in hand, head home from a day of fishing on the lakefront. Nearby is a jogger who has reached the shore through a new park which spans the lakefront freeway, linking the downtown business district to the shore.

Such scenes are growing common along the shores of the Great Lakes and St. Lawrence River. Freight docks, warehouses, grain elevators, and shipping channels still occupy a prominent place in many of the region's harbors, and along Great Lakes tributary streams. But increasingly, they are interspersed with new restaurants, shops, condominiums (often with attached boat slips instead of car garages), and offices touting a lake or river view.

Waterfront parks are also becoming popular. Residents of many lakefront towns can now enjoy miles of waterfront boardwalks and trails, often linked to shoreline parks and nature areas as well as shops, marinas, and other developments.

Recent Great Lakes/St. Lawrence River waterfront construction. More than 194 different waterfront development projects have been planned, permitted, or constructed on the shores of 81 of the Great Lakes' and St. Lawrence River's largest cities since 1986. More than 3,140 acres (1,267 hectares) have been renovated during this period.

These projects have made the Lakes and St. Lawrence River more accessible for recreation, adding more than 13,000 boat slips, 475 acres (192 hectares) of urban waterfront parks and 37 miles (60 kilometres) of waterfront footpaths and promenades. They have made the Lakes and River a better place to visit, adding 2,400 hotel rooms and 1.3 million square feet (387,000 square metres) of shops and restaurants to serve tourists, and a better place to do business, adding more than 5.5 million square feet (1.6 million square metres) of corporate office space with views unmatched by cities in other regions. And these developments have made the region a better place to live, adding almost 7,330 residential units on the shores of North America's most scenic and enjoyable recreation areas.

Even more development is planned for the region's shorelines. More than 3,350 additional acres (1,356 hectares) are planned for development in the future, representing a total future investment of more than $867 million ($C1.0 billion). On the drawing boards are proposals for almost 9,000 additional boat moorings, 690 more acres (279 hectares) of parks, 3.3 million square feet (1.0 million square metres) of new offices, 950,000 square feet (288,000 square metres) of shops, and more than 10,150 more waterfront homes.

The Center's survey found that 76 percent of Great Lakes cities with new waterfront development had included marinas and/or parks. A few cities had only parkland and marinas, but usually those were merely the base for a broader mix of uses. The Center found that 49 percent of the cities had new shopping areas on their waterfronts, while 46 percent had new condominiums or apartments, and 31 percent had museums, amphitheaters, or other cultural centers. One-fifth of the cites included office space in their developments.

Most communities reported that they were reusing land left vacant by the decline of port activity or by manufacturers or railroads that were moving away from the lakefront. Typical of these communities is Port Huron, Michigan. A cement factory and railroad right-of-way along the St. Clair River there have been redeveloped into the Thomas Edison Parkway and Inn, and 31 new condominiums have replaced old warehouses on the shores of the Black River.

A few projects are built in rehabilitated factories and warehouses, such as the renovated Powerhouse on the banks of Cleveland's Cuyahoga River, or Grand Haven, Michigan's Harbourfront Place, built in an historic piano factory. More typical are brand new buildings, ranging from skyscrapers like Etobicoke, Ontario's soaring Grand Harbour condominiums to modest town homes such as the Chicago River Cottages.

Benefits of waterfront development. Well-planned waterfront renewal brings benefits not just to builders and developers, but to all the stakeholders on the shoreline. Throughout the basin, city officials, residents, and business people are finding that renewal of their waterfronts provides an enduring boost in their community's image and quality of life. That boost translates into new business on the shore and an overall improvement in their community's economic fortunes.

Retailers with businesses near the lake find that waterfront visitors are prolonging their stays in shoreside communities, adding an afternoon to browse in gift shops or enjoy a fish dinner in a local restaurant. "It's the best thing to happen to our town," says a Wisconsin businessman. "The waterfront is the basis for all our future development."

Homeowners in neighborhoods adjoining shoreline parks find the developments boost the image, and property values, of their community. "Waterfront development has improved the whole area," enthused the resident of one shoreside neighborhood. "I love to walk here at night with my family."

The economic effects of this renewal are striking:

- In 1981, Duluth was named one of the ten most distressed cities in the country by the *Wall Street Journal.* Just eight years later, *Money* magazine rated it in the top sixth of the most attractive U.S. cities to live. More than 1.1 million visitors now come to Duluth annually, adding more than $100 million in tourism and convention business to supplement Duluth's other industries.

- Studies for Port Colborne, Ontario's Sugarloaf Harbour marina estimate that by 1995 boating-related spending in Port Colborne will total $6.68 million (in 1985 Canadian dollars), which will generate $15.63 million of economic benefits for the community ($12.56 million of which takes the form of wages). A planned hotel and restaurant are projected to generate an addition $40 million in direct expenditures in their first ten years of operation.

■ Grand Haven, Michigan is enjoying an overall boom in its economy, based on its ability to attract both summertime sun-seekers who are drawn to its beaches and marinas, and small manufacturers who have relocated to the city to enjoy its lakeshore and quality of life. Virtually none of this development existed 20 years ago.

Financing waterfront projects. The value of new construction on the region's waterfronts in the past five years is about $3.3 billion ($C3.9 billion), of which 89 percent comes from private investors. This includes not only investments by developers and financiers, but also contributions from foundations and local fund raising efforts. Cities both large and small have benefitted from private fundraising to support waterfront projects, from Traverse City Michigan's 300-foot-long Boardman Riverwalk to Chicago's $43 million Oceanaria. In addition, several communities have recruited local electric utilities as waterfront development partners who contributed rights-of-way and occasionally arranged loans for key development projects.

Public agencies provided 11 percent of the cost of waterfront construction during this period. They often play a key role in sparking waterfront renewal, particularly in assembling building sites and providing sewer and water improvements. Public agencies are also constructing many of the marinas and public accessways along the shore, though increasingly these facilities are being provided by concessionaires or through exactions from private builders.

Local governments provide a critical source of this public financing. In Quebec and the U.S., this often includes general obligation bonds, or, for marinas, revenue bonds secured by the proceeds from slips and marina businesses. Several communities' waterfront improvements are financed using tax increment districts, in which the new taxes generated by the increasing value of waterfront property are set aside to repay the cost of shoreline improvements. In Ontario, regional conservation authorities, which draw funds from both municipalities and from the provincial treasury, have played a key role in many projects. And what community wouldn't envy Manistee, Michigan, which has created a waterfront trust fund with revenues from a natural gas well drilled on city-owned shoreline property, and gained other waterfront development revenue from the disposal of surplus city lands along the shore? Taken together, cities have provided three percent of the financing for waterfront projects since 1986.

Cleveland's Cuyahoga River is the focus of the city's hottest entertainment district, with new developments such as Nautica and Shooter's restaurant (left) becoming a mecca for boaters.
OHIO EPA

Provincial and state grants are the most important public support for waterfront projects, except in Quebec. These include grants for marina construction, often financed with boat registration fees and taxes on marine fuel, as well as various urban planning and economic development funds. All told, provincial and state grants have provided almost five percent of the capital invested in waterfront construction during the past five years.

    Grants from state coastal zone management programs, while modest, often play an important role in preparation of waterfront renewal plans or the construction of waterfront accessways. In Michigan, a unique state trust fund financed with revenues from natural gas wells on state lands plays an important role in shoreline renewal. Ohio's economic development agency has guaranteed loans for some waterfront developments, while New York has designated one key waterfront area as a special "economic development zone", providing property tax abatements, investment tax credits, and other incentives to developers on the site. State transportation agencies' highway programs are playing key roles in reshaping the shoreline at Erie, Pennsylvania, and Duluth. Ontario focuses much of its assistance on programs operated by the ministries of Municipal Affairs and Recreation and Tourism.

    Federal funds, while declining, continue to play a role in waterfront development, especially in Canada. In Ontario, the federal Department of Fisheries and Oceans' Small Craft Harbours Branch financed major portions of the marina projects built in the province, while in Quebec, federal crown corporations and grants from federal agencies are financing much of the new construction on the waterfronts of Montreal, Quebec City, and Trois-Rivieres. Canadian federal agencies also play a key role through Transport Canada's ownership of many port sites suitable for redevelopment.

    In the U.S., important federal programs include Department of Housing and Urban Development Community Development Block Grants and, for marina development, Wallop-Breaux and Dingle-Johnson fishery enhancement funds provided by the U.S. Fish and Wildlife Service. A variety of economic development programs also played a part in financing projects, including the Community Futures Program of Canada's Department of Employment and Immigration, and the U.S. Small Business Administration, Job Training Partnership Act, and Comprehensive Employment and Training Act (CETA). Very few projects took advantage of the incentives allowed in the U.S. for rehabilitating structures on the National Register of Historic Places. Several formerly important sources of U.S. federal funds, such as Urban Development Action Grants and the Economic Development Administration, have been phased out in recent years. Overall, federal programs financed three percent of the cost of waterfront developments in the region.

WATER WORKS

Six Case Studies

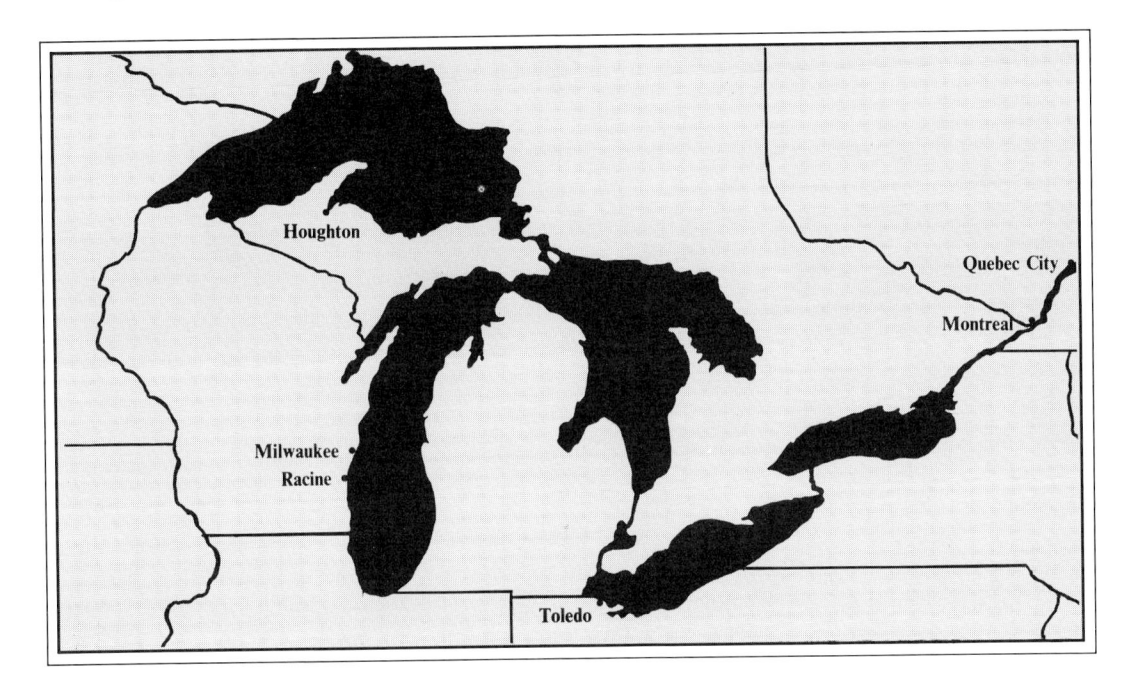

Milwaukee and Montreal—two waterfront development winners. Montreal and Milwaukee, hundreds of miles apart but linked by the Great Lakes/St. Lawrence River system, are each remaking themselves in a fundamental way: they are among scores of cities in the region quite literally turning around to embrace their waterfronts.

The waterfronts were once the very reason for Milwaukee, Montreal, and many other Great Lakes/St. Lawrence cities to spring up and thrive where they did, and have remained central to the commercial life of the region. Now, the waterfronts have gained a new role in the economic life of the region, revitalizing central cities and bringing new investment and new excitement.

"Our water gives us an advantage," said Milwaukee Mayor John Norquist. Milwaukee officials have adopted a strategy of integrating public access to the water with commercial and cultural developments. The city's efforts have focused on both the Lake Michigan shoreline and the Milwaukee River, which flows through the city's downtown.

Milwaukee's lakefront, physically removed from most downtown office buildings, features a lakewalk promenade, and a large park where the city's spectacularly successful Summerfest is held each year, drawing hundreds of thousands of visitors. The newest addition is a Milwaukee County project to turn vacant land near the lakefront into a park atop an underground parking garage. This project will "create direct lakefront/downtown access," City Marketing Director Kris Martinsek said, by linking the lakefront parkland with the end of Wisconsin Avenue, the downtown's main street. Away from the downtown, public beaches encourage recreation along the shore.

The city last year passed a planning guide for the Milwaukee River, which includes four

A system of riverwalks is gradually being created through the heart of downtown Milwaukee.
CITY OF MILWAUKEE

general efforts to improve public access to the river. One, Martinsek said, is the riverwalk system. "Any new developers are strongly encouraged to build riverwalks" of certain minimum widths, with lighting. Another key principle, she said, is to "get developers to design toward the river, and to discourage expansion of non-water-related, non-public businesses along the water."

Milwaukee Center, a public/private mixed-use development covering an entire block along the river, represents a successful culmination of those ideas. The project includes a 28-story office tower, a hotel, retail space, underground parking, and the renovation of a former power plant into a theatre complex for the Milwaukee Repertory Theatre. An enclosed pedestrian concourse links the buildings with each other and with a riverwalk along the water.

The city is also, Martinsek said, "proposing a system of barges with restaurants and concession stands and so forth" to be permanently moored in the river during the warm weather months.

Lastly, Martinsek said, the city is "working on an idea for a water transportation system—water taxis," since the river forms a natural transportation corridor through the downtown. Such a transportation system could, Martinsek said, be extended to the Menominee River, which flows west from the Milwaukee River. Long-range plans call for converting former industrial and railroad lands along that river into public parkland and modern industrial parks.

In Montreal, the emphasis is also on reestablishing the city's connection with the waterfront. One of the priorities of Montreal's first-ever comprehensive urban plan is to create access to the water, according to Lily Robert, a Montreal spokeswoman. Mark London, a Montreal official in charge of plans for redevelopment of two city-owned islands in the St. Lawrence River, said the focus of that project is "to turn the islands' orientation to the waterfront." The city's plans for Ile Saint Helene and Ile Notre Dame are being driven by the idea of "creating an open area on the water for the people of Montreal." The former island was first developed, and the latter one created, for the Expo 67 world's fair in 1967; they are connected to the city's mass transit system.

"We'll be looking at re-naturalizing some of the islands' shorelines," London said, "and building walkways along the river. This is to be a public area, a great urban park." The plan is to develop the major celebration site for the city, right on the water, with a picture-postcard view of the river and the city.

At the same time, plans for the city's federally-owned and historic Old Port area, at the foot of old Montreal, call for a waterfront that will be open to the public, with a children's museum and green space. An IMAX theatre, waterfront walkways, and a promenade along Rue de la Commune, where the port fronts on Old Montreal, have already been completed. The locks of the historic Lachine Canal are being rehabilitated as an historic site, park, and, potentially, a future boating facility. Moorings for tour boats are provided. Already, the Old Port attracts as many as 1.5 million visitors each year. The federal government's investment of up to $100 million in the waterfront is expected to generate five times this sum from the private sector.

Montreal's new plans are the result of public consultations that stretched over several years. Montreal residents insisted that redevelopment of the shore provide access to the St. Lawrence River. They wanted not just physical access, for walking, watching the river scenery, and aquatic sports, but also public uses in the buildings on the Old Port, and visual access, so that the shoreline provided a "window on the river." At public hearings, residents emphasized the need for the uses of the city's shoreline to complement each other and to enhance development in adjacent neighborhoods of Old Montreal. Existing port uses in the area should be consolidated to improve efficiency and their compatibility with newer uses in the port area, but should be retained, people said. Massive private development of the area was rejected.

Racine—linking shoreline revitalization and economic resurgence in a mid-size city. No town illustrates the role waterfront development plays in revitalizing a community's economy better than Racine, Wisconsin. In the early 1980s, Racine's economy, which had the highest proportion of manufacturing employees in Wisconsin, was badly battered. Plants were closing and moving away. The community's aging port saw little use, and its waterfront was neglected. Local leaders, searching for a way out of their economic doldrums, looked to the Great Lakes that had initially provided the competitive advantage for their city's businesses.

The city and county governments, working together with a not-for-profit development corporation representing Racine's business community, put together a $14 million package of local, state, and federal money to convert the port to a recreational boating marina and waterfront festival park. Their proposal was so attractive that before construction ever began, private interests brought in an additional $11.3 million to finance part of the privately-operated marina. By 1988, the city had a state-of-the-art, 921-slip marina, a 16-acre park, and a festival grounds suited for year-round activity. That was just the beginning.

Plans for Montreal's historic Old Port area call for a waterfront open to the public.
GREATER MONTREAL CONVENTION AND TOURISM BUREAU

Racine-on-the-Lake, which converted Racine's former industrial port to a recreational boating marina, sparked millions of dollars of new investment in the city's waterfront and downtown areas.
ROSS DETTMAN

Over the next several years, more than $96 million in new shops and homes were built in the area surrounding the marina. In 1989, the J. I. Case Corporation, one of the North America's largest farm equipment manufacturers, announced plans to build a new $100 million corporate headquarters on the banks of Racine's Root River. Case's executives said they were attracted by the improved quality of life that the redeveloped waterfront offered the community, and were convinced that the highly skilled employees they needed would also find the community's shoreline recreation areas and scenic views an asset. Company officials were also impressed with the "can-do" skills of the city's officials and business community—skills Racine's leaders had honed in conceiving, financing, promoting, building, and operating their waterfront redevelopment program.

Houghton, Michigan—a small town success story. You need not be a major metropolitan city to benefit from waterfront redevelopment. Houghton, Michigan, a small (population 7,500) town on the banks of the Keweenaw Waterway in Michigan's Upper Peninsula, has shown that by working hard, using local resources creatively, and steadfastly following a long-range development strategy, small towns can capitalize on their shorelines.

The former copper mining town seems an unlikely site to find such success. As the mining industry shrank, it abandoned wharves and buildings along the canal. The town's reputation for cold and snow discouraged potential new residents. Cooperation between townspeople and the students and staff at the Houghton's principal employer, Michigan Tech University, was scant. Things got so bad that in 1971, not a single new building permit was issued in the city limits.

That's when Houghton residents decided they needed to act to revitalize their town. "We had one great asset, and that was the water," said City Administrator Ray Kestner.

The idea was to reclaim the waterway's abandoned shore as a green front lawn at the foot of the downtown business district. Waterfront lands were kept for family-oriented public uses, rather than allowing private sector development. State grants were used to buy land along the canal for parks and a jogging and cross-country skiing trail, complete with vista points, picnic tables, and other amenities. New piers and boardwalks were built atop the pilings of old mining and logging piers. Part of the waterway was kept open in case deep-water shipping revives on the canal in the future.

The seven-block downtown adjoining the waterway became an active, pedestrian-oriented shopping area, stretching from the waterfront up to the university campus. The downtown's solid sandstone buildings were dressed up with flags and flower gardens, helping to draw students from the university through the downtown on their way to the canal's shore. Interior passageways were opened between the stores along the main street, allowing winter-time shoppers to move from business to business without stepping out in the cold. Several downtown buildings were rehabilitated to provide apartments for senior citizens, while others found new retail tenants, thanks in part to strict city zoning that focused commercial business on the city's core. Several new parking lots, financed with federal grants, helped support the downtown's revival.

Houghton provided its share of investment in the rehabilitation through hard work and ingenuity. With a high unemployment rate and a total 1970 assessed valuation of less than $9 million, elaborate financial schemes were not in the cards. Often, the city used the labor of its four-person public works crew as an in-kind match for state or federal grants. Another key was gaining the cooperation of downtown businesses, whose support was critical to the downtown tax increment district which helped finance parts of the city's program. Building a good working relationship with supportive state agencies was also important. An annual tour of other communities with successful waterfronts helped refresh city leaders' supply of waterfront development ideas.

Today, Houghton has been designated as one of eleven "communities of excellence" in Michigan. Its assessed valuation has grown to over $38 million. Twelve to 15 new homes are built in the town each year, many in sites just inland of the waterway's parks. Off-street jogging trails are included in new subdivisions, linking every home to the canal-side recreation areas. Waterfront warehouses are being restored as tourist information centers and museums, and historic homes are being renovated by newcomers to the community. During summer the town attracts regattas of recreational boaters from Duluth and Thunder Bay. In fact, the waterway's recreation areas have become so popular that the city has had to provide a separate recreation area for jet-skiing and sail boarding, to avoid conflicts at the bathing beach it created on the canal.

Two troubled projects—Quebec City and Toledo. Successfully redeveloping an urban waterfront is not necessarily as easy as one-two-three, as planners in Toledo and Quebec City can attest. A look at these cities' experiences can provide some instructive lessons on the importance of grassroots support for waterfront projects, and of not imposing grandiose developments ill-suited to local conditions.

In Quebec, the opening of the St. Lawrence Seaway in 1959 led to the Port of Quebec's moving its facilities out of the city's central waterfront, known as the Old Port. By 1970, the area, right at the foot of the historic city, was virtually deserted.

After more than a decade of sporadic public debate, Canada's federal government, which owns the land, settled on a redevelopment plan timed for Quebec City's 450th birthday celebration in 1984. A federal crown corporation, Le Vieux-Port de Quebec, spent $130 million on a redevelopment focusing on Pointe-a-Carcy, which forms the heart of the Old Port. The effort included renovated wharves, landfill, restoration of two historic buildings, and new construction: two open-air theatres, three new "hangar" buildings for shops and restaurants, and covered walkways connecting them.

While the 1984 civic celebrations were successful, the commercial development then failed. Retail establishments in the hangars soon went bankrupt, leaving the federal government with annual maintenance bills of more than $1 million on the mostly empty structures.

Le Vieux-Port de Quebec, having failed to convince the provincial or city governments to take more than the land and the empty buildings for redevelopment, sought ideas from private developers. The plans announced would cover the area with modern office and retail buildings, parking lots, and luxury condominiums. The construction of the first new building, an eight-story condominium project completed in 1986, sparked a burst of public protest and the formation of the Coalition to Save the Old Port of Quebec, an association of organizations and agencies.

In 1988, the federal government declared a moratorium on further development on Pointe-a-Carcy, and in 1989, a federal Advisory Committee on the Future of Pointe-a-Carcy was formed. The committee held public hearings, conducted a public opinion poll, solicited proposals, and late that year, published its recommendations.

Part of the committee's task, said chairman Gilles Boulet in a 1990 speech on Pointe-a-

Carcy, was to assess what had gone wrong with the expensive marketplace buildings and walkways built on Pointe-a-Carcy.

In its report, the committee pointed out that the modernist buildings constructed for the 1984 celebrations clashed with the historic architecture of both the restored buildings and of the nearby "Old Town" area of Quebec City, a popular tourist attraction. "In developing Pointe-a-Carcy, a hodgepodge approach has been adopted, whereas continuity with the Old Town would have been quite obviously much simpler and more felicitous," the committee wrote. "(We are) inclined to believe, in common with some of the participants at the public hearings, that this course was prompted by indulgence of architects' pipe dreams."

Boulet noted that "at the outset, each of the three members of the committee favored the construction of condominiums or offices at Pointe-a-Carcy." But, he said, public testimony and "the constant contact, each weekend, with the crowd" of sightseers flocking to Pointe-a-Carcy "finally convinced us" that Le Vieux-Port de Quebec "had completely missed the mark with its modern buildings and its theatre, and the idea of replacing those buildings with a compact group of housing units, office buildings, hotels and businesses was nonsense."

The committee members, Boulet said, realized that some aspects of Pointe-a-Carcy's development had been spectacularly successful: three cruise ship wharves, a marina, and that "the people of Quebec had adopted the site...as an ideal place to relax, walk with the family, admire the river and its ships. In short, the people had re-established their link with the waterway." Pedestrian traffic has reached 1.5 million people per year.

The committee's charge included finding "a level of economic activity consistent with the nature of the surrounding area." The committee produced a plan based on six basic principles, which were summed up in its main recommendation: "That the future of Pointe-a-Carcy be shaped essentially by its threefold vocation as a seaport, public space for strolling, relaxation and contemplation, and historic site."

The committee recommended that two of the three buildings constructed in the early 1980s, along with the connecting walkways, be removed. No further privatization should be contemplated, the committee concluded, and the heart of Pointe-a-Carcy should become an open public park, while the wharves and marina should remain. The committee strongly endorsed a federal proposal for a museum to be established in the restored Customs Building.

The committee's recommendations were widely hailed by the public organizations which had formed the Coalition to Save the Old Port of Quebec. In August 1990, the federal government

Among the criticisms levelled at early 1980s development on Quebec City's Pointe-a-Carcy was that the modernistic new structures clashed with existing historic buildings.
DANIEL K. RAY

Toledo's Portside festival marketplace opened in 1984 with high hopes for reviving the city's waterfront, but never lived up to expectations; in 1990, it was closed.
CENTER FOR THE GREAT LAKES

announced a plan adopting most of the committee's recommendations, including creating the park and museum, and ruling out further privatization. A federal naval reserve training school is to be housed in one "hangar" building, but without the fences, gates, or other barriers normally associated with a military base.

Boulet said he was "very happy" with the government's decision, and emphasized the "very democratic aspect" of the redevelopment plans. The open, public decision process, he said, helped insure that Pointe-a-Carcy, "so rich in memories and so dear to all Quebecers, will have a future that is in keeping with their wishes."

In Toledo, the outlook is not so bright: Portside, a festival marketplace once seen as the key to reviving the city's downtown area and waterfront, has closed.

Portside was built by the Rouse Company of Columbia, Maryland, in 1984 as part of a $325 million burst of new development along the Maumee River, which included office buildings, corporate headquarters, a major hotel, a river promenade, and Portside's dozens of shops and restaurants. Civic leaders hoped Portside could help revive Toledo's downtown, as Rouse Company festival marketplaces had succeeded in Baltimore, Boston and other cities.

Portside, though, hit hard times in the past two years, as a wave of corporate restructuring turned local officials' estimates of a 50 percent surge in downtown employment into a 20 percent drop. Portside never achieved full occupancy, and its management changed hands four times. By early 1990, Portside could not pay either its debt service or its daily operating costs, and the principal lienholder turned management of the project over to the city. City officials finally pulled the plug in August, after the largest remaining tenant announced plans to leave. A committee of city officials was formed in December to study and recommend possibilities for the building's use.

In retrospect, Portside was probably "too much, too soon," as Toledo Mayor Donna Owens told the *Wall Street Journal* in 1990. While festival marketplaces have succeeded elsewhere, experts say, they cannot be simply dropped into any city and expected to instantly fill up with shoppers.

Bill Owen of The Webb Companies, a Lexington, Kentucky firm which managed Portside in 1989, said, "We had trouble getting the bills paid and attracting businesses. Festival marketplaces in cities of less than a million people have problems, and there wasn't the tourism traffic of, say, Norfolk, Virginia," a medium-sized city with a successful marketplace development. A festival marketplace in Flint, Michigan, a city with demographics similar to Toledo's, has also failed, and been turned into a student learning center.

In a November 1990 post-mortem in the *Toledo Blade,* Rouse Company Vice-President Cathy Lickteig said the firm has all but ceased planning festival marketplaces, because "there are no more cities where the right factors are present. You need the mass, the downtown population." Another development expert told the *Blade,* "When [marketplaces] are not in tourist areas, like Toledo's situation, they just don't work well."

A failure to take into account the character of the local market may have also hurt Portside, according to former Lucas County Planning Chief Barry Hersh. "Toledo is not chic. The strategy of what to market might have been too chic, too upscale. They should have had local companies represented, ethnic products and activities."

Toledo officials had not, as of November 1990, announced long-term plans for Portside. Craig Nebbaoski of Society Bank, the lienholder, said, "The bank has offered to release the liens if the city comes up with a sound development plan."

Keys to Successful Waterfront Development

As the region's waterfront developments mature, shoreside residents are beginning to learn more about the kind of lakefront projects they like. Several communities have imposed temporary halts to construction, while plans for lakefront construction are reexamined.

Toronto's Harbourfront, a complex of marinas, lakefront walkways and plazas interspersed with shopping complexes and hotel and condominium towers, is the site for the most notable of these moratoria. Not long ago, Harbourfront was seen as a model for the region. Now it is derided as "greedy development," "cheap," "ugly," and "The Great Wall of Toronto." Similar controversies have erupted from Quebec City to New Buffalo, Michigan.

Here are several keys to successful waterfront development, gleaned from The Center's development survey and the comments of waterfront officials.

Promote public access and recreation. "People want their waterfronts to be clean, green, open, and accessible", says David Crombie, former Toronto mayor and now Commissioner on the Future of Toronto's Waterfront. Planners in communities as diverse as Chicago, Montreal, and Buffalo have reached similar conclusions.

The result is an increased emphasis on retaining the immediate shore for parks, walkways, and vista points, sometimes enhanced by scattered restaurants and specialty shopping areas. Major developments, like office or condominium buildings, are best located inland of this green space, often separated from the shore by a waterfront drive.

Separating major developments from the shore, studies show, does not result in a significant diminution of their value. In fact, an analysis of properties in Chicago revealed that buildings whose lakefronts are graced by public parks have appraisals equal to or even higher than similar developments which butt right up against the lake.

Nor does the type of new development seem to constrain the ability to provide for public use of the shore. Examples of public waterfront walkways, plazas, vista points, and other accessways are found in literally every kind of development on the region's shorelines, from marinas and restaurants to luxury condominiums and corporate headquarters.

Many communities are lighting their waterfront walkways at night to increase shoreline strollers' feeling of safety, a critical element in encouraging use of shoreline areas. Others are taking extra steps to remove barriers to the handicapped on piers, fishing accesses and other shoreline recreation areas.

On waterfronts where existing structures already separate the city from the shore, new projects seek to break through these barriers, restoring the link between the city and the water. In Longueuil, Quebec, for example, a new waterfront park with bicycle paths, a nature area, and marinas has been established along the St. Lawrence River, restoring public access that had been cut off by the construction of the St. Lawrence Seaway and a riverside expressway. In several towns, such as Monroe, Michigan, and Belleville, Ontario, riverwalks built in the back alleys of downtown shops are helping to reorient businesses towards the water, reconnecting the shoreline and the city center.

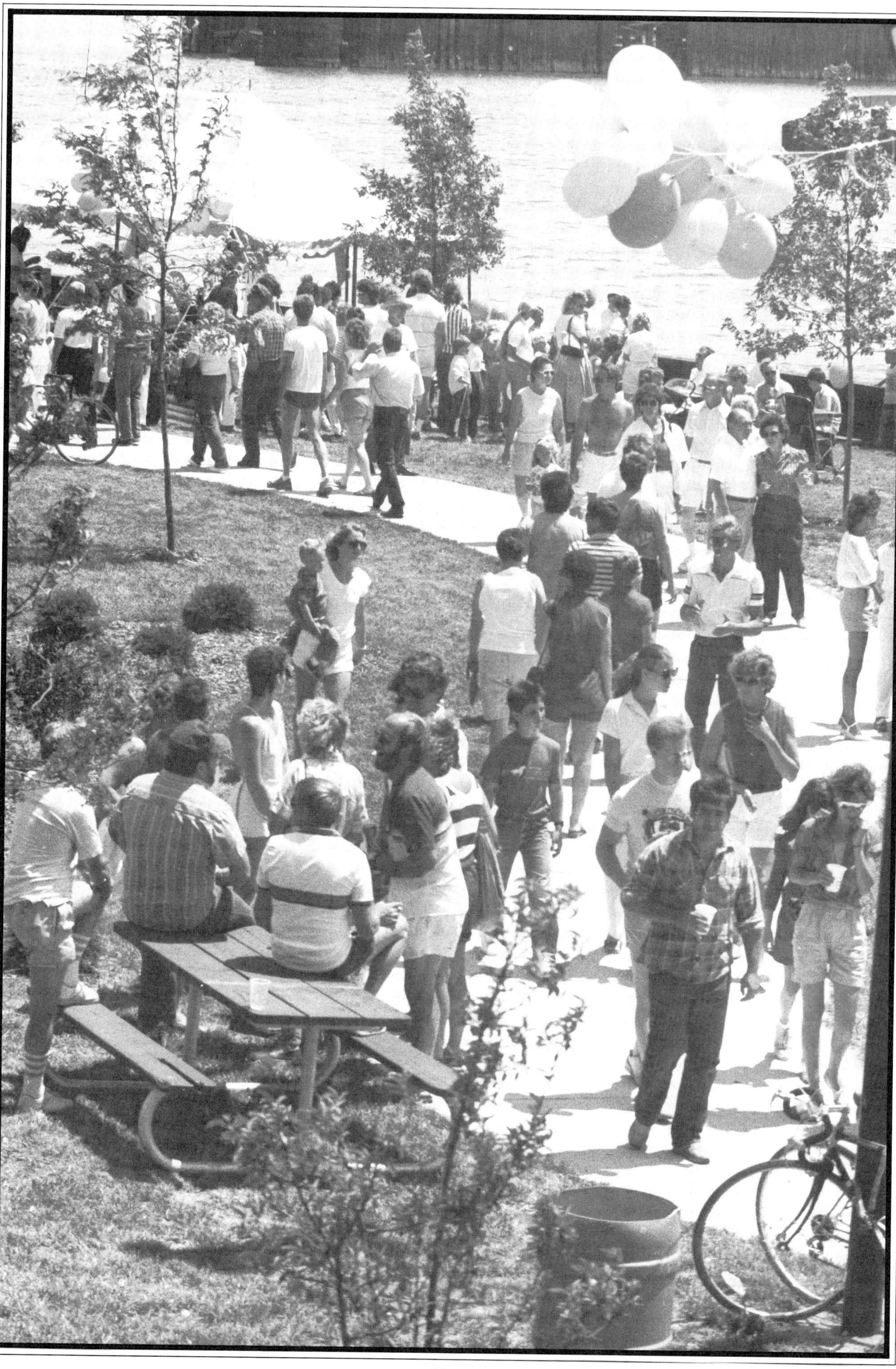

Link waterfront renewal and economic development. Waterfronts are crucial to the economies of cities along the Great Lakes and St. Lawrence River. A 1987 study of Chicago's lakefront suggested that at least 20 percent of the metropolitan area's economic activity "depends, one way or another, on the availability and the attractiveness of the Lake Michigan shoreline."

A waterfront's most important economic effect is often its contribution to a city's quality of life and image. Donna Hinde, an architect with Toronto's M. M. Dillon Ltd. consultants, said waterfronts have become the focus of many cities' overall economic development plans, and not simply because new development can be lured to the waterfront itself. "There's a strong interest now in how the waterfront improvements can improve the image and development climate of the whole community," she said. These long-term strategic benefits to a community frequently exceed the immediate economic value of jobs and investments actually located on the shoreline itself.

Cushman & Wakefield, a major U.S. real estate firm, performs an annual survey of corporate leaders on which U.S. cities are most attractive for locating businesses. Among factors cited as "absolutely essential" for locating office facilities, the percentage of survey respondents listing "quality of life for employees" increased from 18 in 1987, to 23 in 1988, to 32 in 1989. Another study found that quality of life was among ten key attributes that high-technology firms consider when choosing where to locate or expand their business. Surveys conducted for *Money* magazine also show that business people rate access to outdoor recreation and clean water and air as among the most important factors they consider when choosing where to live.

Well-designed Great Lakes/St. Lawrence River waterfronts, with plenty of parks, areas for boating and fishing, and waterfront accessways, provide the recreation and quality of life attributes that these businesses seek. As the population of the region's cities grows older and more sophisticated, the market for these downtown waterfront improvements will increase.

Enhancing tourism is another key role that waterfront development plays in the region's economy. "Anyone who looks at a map of North America sees this huge body of water in the middle," Milwaukee's Mayor John Norquist pointed out. "But our tourism potential has gone largely untapped." Shoreline projects which emphasize the unique heritage of the region, such as Cleveland's Flats entertainment district or Sheboygan's Shantytown fishing village, enhance tourism opportunities in waterfront towns, as do cultural or recreation features, such as Vermillion, Ohio's Great Lakes Historical Museum or the charter sailing center in Superior, Wisconsin's Barker's Island marina. Waterfront festivals, like Rochester's "River Romance" or Windsor's "Freedom Festival," also contribute to the excitement that draws visitors to the shore.

A word about marinas. In most Great Lakes communities, a marina is an important recreational feature. Often, they are viewed as the key to revitalizing the waterfront and increasing shoreline tourism and employment. More than 65 marinas have been built or expanded in the region since 1986, taking advantage of the boom in recreational boating brought on by restored fisheries and water quality. New development planned for the region would continue to add boat slips at a comparable pace.

Increasingly common are marina facilities to serve tourists and non-boat owners, as well as moorings for local residents' boats. Several marinas have consolidated sportfishing docks, special charter-sailing centers, and extra-large berths for tour boats, allowing those without their own vessels to enjoy boating on the Great Lakes and St. Lawrence River.

Berths reserved for transient boaters, marina laundromats, and other features can help marinas attract the growing number of boaters cruising amongst the Lakes' harbors and islands. Pump-out stations to protect the Lakes' water quality, as well as slips and other facilities for marine patrols and law enforcement, are also essential. Harbor experts are encouraging these kinds of facilities, noting that waterfront designers should be as attentive to how well their developments serve sailors, cruise operators, charter boat captains, and other boaters as they are to the needs of landside day-users, retailers, and waterfront homeowners.

Builders planning new marinas need to carefully study the future market for boating. Reports on the uncertain future of the Lakes' sport fisheries, concerns about over-building pleasure boat capacity, and the high cost of marina maintenance underscore the need for careful planning of new marina projects. Declining sport fisheries in Lake Michigan (and potentially Lake Ontario) and the added marina maintenance cost of controlling zebra mussels, a foreign organism

Waterfront developments succeed most when they promote public access to the water, as in Manitowoc, Wisconsin's lakefront walkway.
RON HOERTH

now infesting the Great Lakes, are just two of the many questions which must be addressed in assessing new marinas' financial feasibility.

Smaller communities considering marina construction for the first time should also evaluate carefully their long-term capacity to manage and maintain complex boating facilities. To succeed, a marina needs professional managers who can provide aggressive marketing to attract boaters, quality service to retain them, and shrewd business skills to deal with concessionaires operating shops and restaurants. Marina operations must also be capable of not only repaying interest on construction loans, but also generating revenue for long-term maintenance and periodic improvements.

Of particular concern are marinas where the slips are assigned to individual owners though long-term leases (some up to 99 years) or "dockominiums." These new kinds of marinas will tie up shoreline areas for years to come, and have an unproven record of marina operation and maintenance. Others are concerned that they are privatizing the shoreline in ways that may interfere with public use of the shore. Some communities are limiting this form of marina ownership, requiring a certain percentage of slips be maintained as annual rentals, or restricting dockominium marinas to areas dredged from dry land.

Sustain working waterfronts. The sight of a laker, slowly working its way up Cleveland's Cuyahoga River amidst power boats, jet skis, and other watercraft, provides a unique setting for riverfront diners in the riverside restaurants. For the ship's pilot, small craft rafting out from the restaurants five and six abreast, to join thousands of pleasure boaters squeezed into the navigation channel, complicate his challenge at maneuvering the 630-foot (128-metre) ore carrier through the river's narrow oxbow bends. For the shipowners, steelmakers, and thousands of workers whose fate is tied to the efficiency and safety of commercial navigation, the lost time and added risk that can accompany this and other conflicts between port uses and waterfront developments dull the competitive edge of the region's industries.

Conflicts between waterfront redevelopment and traditional maritime uses of Great Lakes and St. Lawrence River ports arise from a variety of sources: competition for space in navigation channels, harbors, and waterfront lands; congestion along truck routes leading to waterfront mills, wharves, and bulk terminals; and complaints from new waterfront residents and visitors, who may object to the noise, dust, and operations of a working waterfront. If these conflicts become too severe, the port operations whose diversity helps make Great Lakes/St. Lawrence River waterfronts unique (and provide a four-season, high-wage economic base for shoreline communities) may become uneconomical. Loss of these commercial port operations diminishes not just the maritime

Careful planning can avoid conflicts between new recreational uses of a waterfront, and the traditional "working waterfront" still important to many local economies in the region. At Quebec City, walkways are safely separated from the working port while still allowing sightseers a view of maritime activities.
DANIEL K. RAY

traditions and economy of individual communities, but also the critical mass of skilled workers, fleet operators, and other businesses required to sustain Lakes navigation and finance the maintenance of the Seaway and our harbors and channels. Once gone, this waterfront space may be impossible to reclaim for new port uses, like moorage for tour boats, which may develop in the future.

Great Lakes and St. Lawrence River communities are using a variety of tools to protect the maritime uses of their mixed-use waterfronts. These tools begin with comprehensive harbor management plans that identify sites where new waterfront development can locate without interfering with maritime uses. Some towns implement these plans with zoning ordinances and bylaws that restrict waterfront lands for exclusively water-dependent uses, prohibiting residences and incompatible commercial development. Others have ordinances to protect existing waterfront uses from nuisance suits brought by new residents and recreation businesses. In Cleveland's Flats Oxbow district, port industries join architects, waterfront users, and other members of a design review committee to examine projects for features which may interfere with neighboring port or industrial uses.

Maintaining navigational use of harbor waters is critical to protecting working waterfronts. Some harbors have established safety zones which regulate pleasure boat moorage to protect navigation channels, wharves, and turning basins for maritime use. Special design features can help, too, like bulkheads which discourage mooring of pleasure boats in unsafe locations.

In some working waterfronts, harbor authorities have turned to land banking, in which ports purchase and set aside sites for future port expansion. This may include relocating some port facilities from the downtown waterfront to more peripheral locations. Others have subsidized port facilities with revenues from compatible development of underutilized waterfront areas. Some state or provincial agencies offer low cost loans to finance expansion or improvement of port uses.

Helping visitors and townspeople understand the needs and importance of waterfront industries is the key to any of these approaches. Most successful approaches begin with multi-stakeholder waterfront committees or associations, where port operators, shippers, waterfront developers, residents and others can share concerns, devise their own strategies for avoiding conflicts, and educate city officials about their needs. Vista points that allow residents and visitors to view working ports safely can also help build this understanding, as can signage that interprets intrinsically fascinating maritime activities. Other approaches include festivals and development themes that acclaim the working waterfront, including activities like tug boat races and factory tours, and outdoor art that celebrates shipping and shoreline industry.

No one tool can protect all the maritime uses of our urban waterfronts. After all, it is partly the decline of waterfront industries and ports that makes waterfront redevelopment possible. Nostalgia for the flush times of the past will not keep port uses thriving in the absence of real market demand. Where economic conditions are favorable, communities with a shared vision of the strength of their waterway can keep a waterfront that works.

Restore waterfront environments. The success of the region's waterfront redevelopment efforts is inextricably linked to progress in restoring the Lakes and St. Lawrence River themselves. Sandusky would not have built more than 2,600 boat slips in the last five years if Lake Erie were still choked with algae and its walleyes still contaminated with mercury, nor would more than $700 million be committed and planned for waterfront development in Cleveland if the Cuyahoga River were still flammable.

According to real estate and development specialists at the Urban Land Institute, "one of the most significant factors stimulating [waterfront development] is the dramatic improvement in environmental quality." They go on to say "an effective urban waterfront renewal plan must meet a critical precondition—clean water. Without clean water, not even the most innovative and appealing project will succeed in attracting people and activity to the banks of a river or bay."

Nowhere is this fact more apparent than in the restored waters and revitalized waterfronts of the St. Lawrence River and the Great Lakes. The roughly $11 billion invested in municipal wastewater treatment plants in the region since 1970 has paid off in a multi-billion dollar sportfishing industry, booming growth in recreational boating, and a return of visitors and residents alike to the region's shorelines. But the work of restoring the Great Lakes and St. Lawrence River is incomplete, especially in the 66 persistent pollution hotspots which have been labelled as Areas of Concern under the Canada-U.S. Great Lakes Water Quality Agreement and, in Quebec, as zones

Improvements in the water quality at Areas of Concern like the Maumee River is critical to the success of new development along the Lakes, including Toledo's downtown area. TOLEDO-LUCAS COUNTY PORT AUTHORITY

d'interet prioritaire in Environment Canada's St. Lawrence Action Plan. These areas include the harbors, riverfronts, and nearshore areas of many of the region's cities.

Completing the restoration of water quality in these areas is essential to the success of waterfront revitalization on the Lakes and River. Seventy percent of the waterfront development experts surveyed by The Center agreed that improving water quality would increase the value of waterfront property and developments in their area. One in four could point out specific waterfront developments that had been delayed or scaled down due to water quality problems. People who use the waterfront concur with these experts. Users of Milwaukee's Riverwalk, for example, said the riverfront improvement most needed was a cleaner river.

The need to restore these areas of degraded water quality is beginning to be recognized as a key part of the region's waterfront revitalization efforts. In Belleville and Thunder Bay, Ontario for example, toxic cleanups and habitat restoration projects for local Areas of Concern are included in the cities' waterfront development programs. In Oswego, New York and Milwaukee, city officials coordinated the development of wastewater and waterfront projects, saving construction costs and making innovative use of construction rights-of-way and materials.

Waterfront builders are well informed about water quality issues, with more than 80 percent of those surveyed by The Center aware of local water quality problems. Yet, despite the high stakes which waterfront developers have in the restoration of the Lakes and River, more than half the waterfront developers surveyed by The Center were not participating in the remedial action plans for their waterfront's Area of Concern, even through a trade association or other group. More needs to be done to link these waterfront builders with the restoration of the waters on which their livelihoods depend.

More effort is also needed to protect natural areas along the region's shores. The Center's survey found that nine miles (14 kilometres) of shoreline were armored, paved, or bulkheaded in the region for waterfront development since 1986. Almost 800 acres (320 hectares) of the Lakes were filled to provide building sites. These alterations, while sometimes unavoidable, eliminate fish and wildlife habitat that draw people to downtown shorelines. In addition, poorly designed shoreline modifications can block the movement of sand along the shore, causing erosion of nearby beaches. In many Great Lakes/St. Lawrence River cities, this new waterfront development could instead be located on abandoned industrial or port land where natural shorelines would not be damaged.

A few Great Lakes communities are beginning to recognize that the benefits of protecting and restoring these natural shorelines may offer more to their towns' quality of life and image than they could contribute as development sites. In Traverse City, for example, a beach has been restored on abandoned bayshore railway lands. In metropolitan Toronto, the regional conservation authority is creating artificial gravel beaches out of concrete rubble and rock left over from construction projects. In Montreal, natural shorelines are being restored on islands in the St. Lawrence River. Others are finding ways to enhance fisheries during waterfront projects, using clean rubble from demolition to create offshore reefs, or to restore or create wildlife habitats to offset damage from developments.

Cleaning up development sites contaminated with toxic chemicals remains a challenge for communities along the Lakes and River. In Sheboygan, Bay City, Michigan, Toledo, Buffalo, and Hamilton, Ontario contamination of harbor sediments and old industrial sites is preventing construction of marinas or other projects, or increasing the cost of development that is occurring. Some sites are too contaminated to allow any kind of development, creating holes in the fabric of waterfront renovation. New ways of overcoming the barriers created by contamination are needed, through innovative cleanup technologies and alternative uses for contaminated sites.

Celebrate winter. Skaters clad in bright coats, scarves and hats glide past dragons and pagodas sculpted in ice. Shoppers in a waterfront galleria gaze out from a glass observation deck at the fury of a winter storm. Tourists sip hot buttered rum while waiting for horse-drawn sleigh rides at a shoreline festival site ablaze in colorful pennants and bright lights. These are the pictures of winter on Great Lakes and St. Lawrence River waterfronts.

"Lake effect" snows—lakeshore snowfalls up to 20 percent heavier than those of inland communities—favor these developments. Across the basin, builders are beginning to find ways to use the shoreline's moist microclimate as an advantage in luring visitors to the waterfront in winter as well as summer. Careful attention to design details helps these waterfronts blunt winter winds which can sweep into the city from the warmer, open Lakes. Other features assist in snow removal, control damage from ice, and make the sites attractive for year-round use.

Successful year-round use of waterfronts begins with a good understanding of how winter will affect a project. A wide, sunlit plaza that could be wonderful in a sunny, Mediterranean waterfront, for example, would likely be icy, windswept, and abandoned during a winter day on the Lakes. Instead, a waterfront project here would use shrubs, screens, canopies, berms, and other barriers to break up the site and dissipate the wind.

Changes in building design also help. Placing the narrow sides of a building's corner so it faces the prevailing wind, employing a stepped design for building facades, and using round corners instead of sharp ones can reduce drafts and turbulence. For large projects, a mathematical model and close work with an aerodynamic engineer can be used to test how proposed buildings will perform in winter gusts.

Encouraging exposure to the sun is important. Tall buildings should be placed where they will not cast winter shadows over shoreline accessways and gathering spots. Galleries, winter gardens

Each winter, an ice fishing village is built on the frozen St. Anne River near its outlet on the St. Lawrence River at St. Anne de la Perade, Quebec.
ST. ANN'S FISHERMEN ASSOCIATION

and other greenhouse technologies can help take advantage of the season's shorter periods of natural sunlight. Pastel colors and bright lights can add splashes of color to the flat winter landscape and offset the longer hours of darkness.

Evergreens can provide landscaping that is as attractive in January as June. New designs also permit year-round operation of fountains and other water features. Enclosed observation areas allow visitors to enjoy these natural features without sacrificing warmth and comfort.

Other design innovations make it easier to use waterfront developments despite the season's snow and ice. Summertime reflecting pools and patios are being designed to serve double duty as winter skating rinks. Some projects have snow-thawing systems, sometimes channelling heated wastewater from local industries through subsurface pipes to warm sidewalks and patios. Others have raised walkways and curbs that protect pedestrians from spring slush and splashed, muddy water. Skywalks and covered arcades also make getting about on the waterfront easier in winter.

Marinas include features to offset harsh winter conditions, too. Some use compressed air bubblers to suppress ice that otherwise could damage piers and pilings. The open water that results may draw geese or other wildlife to be viewed by visitors to the development. Special paints and insulation are among new products designed to slow icing of pilings. In well-protected harbors, new designs allow floating docks to freeze in and melt out with few, if any, problems. Launching ramps that can serve ice boats and ice fishermen as well as watercraft can provide four season use of rivers, canals and coastal lakes.

Winter festivals keep the waterfront alive year-round. Some are major regional events, like Quebec City's famed Winter Carnival, Niagara Falls' Festival of Lights, or Milwaukee's Winterfest. Others, like Akron's midwinter Chili Open golf tournament or the village of ice-fishing shacks erected at the St. Lawrence River town of St. Anne de la Perade, are local attractions. Winter sports, such as cross-country ski races, dog sled contests, and snow shoe races, add to the waterfront's excitement.

The key is a waterfront with a mix of facilities to generate year-round activity, but flexible enough to accommodate fluctuations in demand for certain use. What results is a new style of shoreline development that celebrates winter, showing the world that cold weather on the Great Lakes and St. Lawrence River can be fun.

Build regional partnerships. Another trend in the region is the increasing emphasis on coordinating development among neighboring waterfront communities, so that their waterfront projects complement each other. In this way, efforts in adjoining communities strengthen each other, rather than competing for the same market.

In Indiana, for example, the Northwest Indiana Regional Planning Commission has begun a study of regional waterfront development priorities. In Ontario The Royal Commission on the Future of Toronto's Waterfront is calling for "partnership agreements" between waterfront cities and the province, helping to enhance the overall compatibility of developments in adjoining communities. In western New York, eight cities and Erie County are collaborating in an overall waterfront redevelopment program through the Horizons Waterfront Commission.

At the regional level, Ontario and the Council of Great Lakes Governors have launched a joint effort, "The Fresh Coast," to market the region's lakefront tourism opportunities, and Quebec, Ontario, and the Great Lakes Commission are collaborating to promote a "Great Lakes Circle Tour" motor route.

Be yourself. Those communities whose shoreline developments are most successful have found a way to build a bit of their own unique style into their waterfronts.

Manitowoc, Wisconsin's waterfront features a maritime museum that celebrates the city's history as a boat building center. St. Catharines, Ontario's revitalization of Port Dalhousie builds on the historic character of old neighborhoods along the Welland Canal.

Cleveland's Flats district integrates nightclubs and shops with manufacturing and shipping operations along the Cuyahoga River. Toronto and Detroit feature the arts in their shore areas.

The key is to promote diversity in shoreline development, and to emphasize the unique attributes of each community in its waterfront projects, rather than parroting development patterns found in every town from Buffalo to Thunder Bay.

WATER WORKS

City Profiles

The 81 Great Lakes/St. Lawrence River communities profiled in this survey, while but a sampling of waterfront development in the basin, provide detailed and varied examples of the revitalization of the shorelines and economies of the region's cities. They also reflect a deepening appreciation of the value of the Great Lakes and St. Lawrence River themselves.

The information presented on the following pages was gathered in 1990 in response to a telephone survey administered to medium-sized and large communities. Completed profiles were reviewed by city officials and development experts to verify their accuracy. The cities and projects profiled reflect the full diversity of waterfront redevelopment underway in the basin, and are not intended to endorse specific projects or types of development.

The projects listed include only those located between the shore and the first parallel public road. Developments located inland of lakefront parks and waterfront drives were not included in the survey. Construction costs and financing information are presented in the currency of each city's nation. The profiles are presented on a lake by lake basis, beginning on Lake Superior and following the flow of the basin's waters to the St. Lawrence River. A summary of development and a list of noteworthy projects are presented for each Lake and the River, followed by city profiles presented in clockwise order. A matrix presented on pages 67 and 68 characterizes each city's developments by uses and the financing mechanisms used. You are encouraged to contact the individuals listed for additional information.

More detailed information on many of the projects described here, including a complete list of project developers and consultants, is available in Fact Sheets from The Center's Great Lakes Information Service. See page 72 for a form for ordering waterfront development Fact Sheets from The Center.

Chicago leads the region in waterfront development, with a variety of projects such as the Illinois World Trade Center, providing new access and amenities along the Chicago River. PHIL GREER/CHICAGO TRIBUNE

LAKE SUPERIOR
Including the St. Marys River

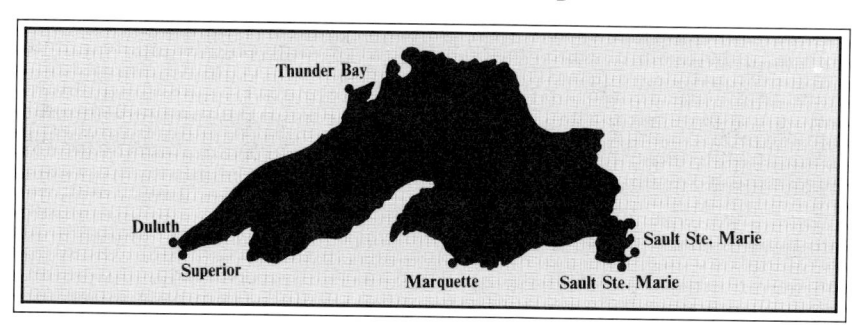

The port cities of Lake Superior's rocky shores are finding a new niche as centers of tourism and the home bases for a thriving charter sailing industry.

More than 224 acres (91 hectares) have been redeveloped since 1986 in the cities surveyed by The Center. More than $59 million ($C68 million) has been invested in new construction, resulting in more than 20,000 square feet (1,800 square metres) of new shops, 146 new marina slips, 60 new homes, 25 acres (10 hectares) of new or rehabilitated parks, and 1.75 miles (2.8 kilometres) of new shoreline walkways. Proposed for future development is an additional $18.8 million ($C22 million) in projects, including almost 150 more boat slips, 150 homes, five acres (two hectares) of parks, and two miles (3.2 kilometres) of shoreline paths.

Among the highlights of Lake Superior's waterfront development are:

- The neighboring Ontario and Michigan cities at Sault St. Marie, who are collaborating in the development of tourist facilities focused on the Soo locks.
- Duluth, which is mid-way through an ambitious waterfront revitalization.
- Thunder Bay, whose recent waterfront development plan links shoreline redevelopment and the restoration of the area's water quality.

DULUTH, MINNESOTA

Duluth is continuing the ambitious downtown waterfront development program it first adopted in 1985. More than a mile of the Downtown Lakewalk, a pedestrian boardwalk and paved trail for bicycles and horse carriages along the Lake Superior shore, has been completed. The Lakewalk includes a .6-acre lakefill, constructed with blast rock from the construction of Interstate 35. Gravelly Newfound Beach formed at an artificial headland constructed as part of the project. A final phase of the Lakewalk, scheduled for completion in 1991, will link the two completed sections to create a 2.5-mile trail along the city's North Shore, from Canal Park to Leif Erickson Park.

An extension of the Lakewalk west of Leif Erickson Park is being constructed by the Minnesota Department of Transportation as part of its improvement of Interstate 35. Also under construction in this area is Lake Park Place, a 3.5-acre public park atop the right-of-way of Interstate 35. The Lakewalk shares a right-of-way through much of its length with the St. Louis and Lake County Rail Authority, which operates an excursion train service from Duluth to Two Harbors.

The $1.4 million cost of the Lakewalk was financed from the city's general fund. Lake Park Place and the trail extension past Leif Erickson Park are being built and financed as part of the completion of the Interstate 35 extension.

In Canal Park, located alongside the ship channel linking Duluth Harbor and Lake Superior, Canal Park Drive is being renovated. The road's traffic lanes are being narrowed, and the sidewalks are being widened and bricked. Street furniture, sculptures, and a tower marking the entry to the downtown waterfront district are also being added. The $7.2 million project, financed with municipal general obligation bonds and a grant from the Minnesota State Highway Fund, will be completed in 1991. Recent retail projects in the Canal Park neighborhood include Grandma's Sports Garden, a new restaurant, and expansions of Grandma's Saloon and Deli and Mickey's Grill and Shops.

Also proposed for Canal Park is Lake Superior Center, an interpretive, research and meeting center planned to include an aquarium. A preview exhibit will open in 1991 in

Bayfront Park on Duluth Harbor. Financing for the multi-million dollar facility is expected to come from the Duluth area, state funds, and private, foundation, corporate, and federal funds. The project is being planned by a not-for-profit organization, which is cooperating with scientists from the Soviet Union to develop common programs to study large lakes.

Revitalization of the waterfront is also continuing inside Duluth Harbor. The new Waterfront Plaza Charter Fishing Dock provides berths for 30 charter boats at the Minnesota Slip, an old wharf. The nearby Duluth Entertainment and Convention Center has added a $17.2 million conference center, exhibition space, and ballroom, financed with state general obligation bonds. It was planned and built by a special-purpose authority with a governing board representing both the state and city. On 16 acres west of the convention center, a 140-slip marina and 80 to 150 condominiums are planned.

CONTACTS:

GERALD M. KIMBALL
City of Duluth Planning Division
City Hall, Room 409
Duluth, Minnesota 55802-1197
(218) 723-3328

DAN RUSSELL, Executive Director
Duluth Entertainment and Convention Center
350 Harbor Drive
Duluth, Minnesota 55802
(218) 722-5573

ROBERT J. BRUCE, Executive Vice-President
Lake Superior Center
353 Harbor Drive
Duluth, Minnesota 55802
(218) 720-3033

DAVID DAVIDSON
Minnesota Department of Transportation
1123 Mesabi Avenue
Duluth, Minnesota 55811
(218) 723-4842

SUPERIOR, WISCONSIN

The Barker's Island Marina in Superior is among the largest marinas on Lake Superior. CITY OF SUPERIOR

In 1988, 60 slips were added to the city's 360-berth Barker's Island Marina on Superior Bay. The $220,000 development was funded by city general obligation bonds and a grant from Wisconsin's coastal zone management program. Six condominiums were built near the marina, the first phase of the Barker's Island Townhomes.

CONTACT:

MARSHALL WEEMS, Director of Planning and Development
City of Superior
1407 Hammond Avenue
Superior, Wisconsin 54880
(715) 394-0335

THUNDER BAY, ONTARIO

This historic port city recently completed a plan for the revitalization of its South Core waterfront on the Kaministikwia, Mission, and McKellar Rivers. The plan's key element is development of Kaministikwia Heritage Park. Its central plaza would have a carousel, restaurants, heritage displays, shops, an amphitheater, and a shoreline footpath. Boat mooring would be allowed along the park's shoreline bulkhead. Winter facilities include a skating rink (which will double as a reflecting pool in the summer).

Improvement of nature study and access facilities at the Mission Marsh Conservation Area is also planned. Improvements could include a visitor center, boat dock, footpaths, and picnic areas. Other planned waterfront improvements include two new neighborhood parks, 110th Street Park and Mountdale Avenue Park on the Kaministikwia River, and enhancement of Current River Park, Floodway Lookout Park, and Kam River Woodlot. A comprehensive accessway signage program is also proposed.

Other elements of the plan include designating a site for a proposed private marina on Mission Island on the McKellar River, and clean up of toxic hotspots on the Kaministikwia, Current, and McKellar Rivers, the Neebing-McIntyre Floodway, and McVicar Creek. These cleanups are expected to cost about $1.4 million. The cost of the $15.3-$17.9 million project is expected to be shared by the city, the province (including funds from the Ontario Ministry of Municipal Affairs' PRIDE program [Program for Renewal, Improvement, Development, and Economic Revitalization] and the ministries of the Environment and Northern Development and Mines), Environment Canada's Great Lakes Clean-Up Fund, and a private marina developer.

Recently proposed improvements in Thunder Bay's northern waterfront include the addition of 75 slips to the Prince Arthur's Landing marina.

CONTACT:

PAUL FAYRICK
Planning and Policy Development
City of Thunder Bay Parks and Recreation
950 Memorial Avenue
Thunder Bay, Ontario P7B 4A2
(807) 625-2806

SAULT STE. MARIE, ONTARIO

The city, located at locks which connect Lakes Superior and Huron, is diversifying its battered manufacturing economy with tourist-oriented waterfront development. A central feature of the city's plan is relocation of a scrap yard at the terminus of the International Bridge across the St. Mary's River to make way for a major tourist attraction, the True North Experience, including an historic exhibit, Omnimax theatre, and indoor recreational ride (perhaps simulating a canoe ride on the St. Marys River rapids). A tourist information centre, Great Lakes water park, and a $4 million freshwater aquarium are also planned for the site.

Already under construction is 2.4-hectare Norgoma Marine Park, a new 56-slip marina and a berth for the *Norgoma,* a 55-metre former Great Lakes passenger steamship now operated as a museum. An open-air pavilion sheltering up to 400 people for festivals, markets, and other events will be added to the park in 1991, along with a marine service building (fuel sales and a pump out) and a lighted riverfront walkway. Planned for the future are a reflecting pool that can be used as a skating rink in winter, and a pedestrian ferry crossing the river to Sault Ste. Marie, Michigan.

A festival site with a 1,000-seat outdoor amphitheater is planned for Clergue Park, at the city's waterfront civic centre. To facilitate this development, the city has relocated a sand and gravel operator to a 20-hectare, riverside site outside of the downtown, and demolished two oil and gas tank farms. A bush pilot and forest fire museum will be developed in the riverside hangars formerly occupied by the sea plane base.

To link these developments, the city is constructing a continuous pedestrian and bicycle path along the river. A pedestrian bridge with a fishing access has been built at the municipal fish hatchery, and nearly two kilometres of the riverside pathway have already been constructed. The city will rebuild the walkway in front of the Holiday Inn. The trail is groomed for cross country skiing in winter.

Sault St. Marie expects to spend $4 million, funded by provincial grants, on its waterfront development program in 1991.

CONTACT:

JOHN BAIN, Planning Director
City of Sault St. Marie
Civic Centre
P.O. Box 580
Sault St. Marie, Ontario P6A 5N1
(705) 759-5368

SAULT STE. MARIE, MICHIGAN

The "Soo", like its Canadian neighbor across the St. Marys River, focuses its waterfront development on the busy locks in its downtown. Its centerpiece is the Locks Park Historic Walkway, a mile-long handicapped-accessible footpath along the St. Marys River and the Soo Locks. The $50,000 walkway was financed by the city and a grant of the Michigan Department of Natural Resources' coastal zone management program. Because the walkway includes land leased from the U.S. Army Corps of Engineers, the city had to provide extensive insurance to cover any liability arising from public use of the area. Public use of the riverwalk has not adversely affected the lock's operation.

Other recent improvements in the area include construction of new berths for the Famous Soo Locks Boat Tours ($500,000, privately financed) and the Pitsch Company apartment complex, with 60 apartments and shops ($3.8 million project, privately financed). The apartment site, a tax delinquent property held by the Michigan Department of Natural Resources and two adjoining parcels owned by the city, was made available to the builder at a fixed acquisition cost. Other recent developments in Sault St. Marie include the Ashman Bay Launching Ramp and improvement of a swimming beach, 20-space campground, picnic grove, and a habitat conservation area at Sherman Park. Both projects are on the Upper St. Marys River. The two projects cost $250,000, with financing from the city, the Michigan Land and Water Conservation Fund, and the Edison Sault Electric Company, from whom the city leases the boat launch site.

CONTACT:

JIM HENDRICKS, Director of Planning and Development
City of Sault Ste. Marie
1301 W. Easterday Street
Sault Ste. Marie, Michigan 49783
(906) 635-9131

MARQUETTE, MICHIGAN

A former coal dock here has been redeveloped into 20-acre Ellwood A. Mattson Lower Harbor Park. Four feet of fill was imported to cover a concrete slab on the site, and a quarter-mile shoreline path was built. It ties into nine miles of existing paths on both sides of the park. The dock's breakwater was retained, and a site will be dredged for a proposed marina. The $1 million project is financed by private donors, the city, a federal Community Development Block Grant, and a grant from Michigan's coastal zone management program.

CONTACTS:

NORMAN GRUBER
City of Marquette, Planning Department
300 W. Baraga Avenue
Marquette, Michigan 49855
(906) 228-0430

JOHN TURAUSKY
Marquette Park Department
300 W. Baraga Avenue
Marquette, Michigan 49855
(906) 228-0460

LAKE MICHIGAN

There are few towns on Lake Michigan that have not begun to redevelop their waterfronts. On the Lake's rural northern coasts, new sportfishing and tourism developments are underway. In the ports which once served the Lake's mid-sized manufacturing towns, marinas, condominium developments, specialty shopping areas, and corporate offices are replacing abandoned wharves, grain elevators, and factory sites. And in Milwaukee and Chicago, glittering downtown skyscrapers are adding promenades, marinas, parks, and outdoor cafes along the cities' riverfronts.

More than 628 waterfront acres have been developed since 1986 in the 28 Lake Michigan cities surveyed by The Center. More than $2.9 billion in new construction has occurred, including a total of 1,900 hotel rooms, one million square feet of restaurants and shops, 2,000 residences, and 5.3 million square feet of offices. Almost 150 acres of parks have been created or improved along these cities' shorelines, adding more than 4,200 new marina slips and 18 miles of new waterfront footpaths.

There are plans for development of more than 1,300 additional waterfront acres, including more than 380 acres of parks, 500 hotel rooms, 5,200 boat slips, 86,000 square feet of retail space, 1,600 residences, and three million square feet of offices.

Of particular note are:

- Racine, where redevelopment of the city's harbor sparked a broader economic resurgence.
- Milwaukee's Henry W. Maier Festival Park, the region's premier waterfront festival site.
- Sheboygan, whose Shantytown development showcases the region's commercial fishing industry.
- Traverse City, where part of the Grand Traverse Bay waterfront is being restored to more natural conditions.
- Manistee, which is combining funds from many sources to finance its waterfront redevelopment.
- New Buffalo, where marinas and waterfront condominiums have transformed a once sleepy community.
- Indiana, whose Lake Michigan Marina Development Commission is facilitating a major increase in pleasure boat facilities.
- Chicago, home of Cityfront Center, the largest waterfront development in the region.

RACINE, WISCONSIN

Racine's Racine-on-the-Lake development includes a colonnade connecting two event areas, which is used for farmers' markets and art fairs. ROSS DETTMAN

Redevelopment of Racine's harbor has sparked a resurgence of the city's economy. Racine's success began with Racine-on-the-Lake, a 132-acre mixed-use development in the city's former port at the Root River's outlet to Lake Michigan. The project includes a 921-slip marina, a six-lane boat launching ramp, two service buildings, and 450 parking spaces.

An administration building adjoining the marina includes a restaurant and specialty and marine-related shops. The project also includes a festival park with a large event area (for up to 5,000 people) and a small event area (for up to 1,000 people) connected by a covered colonnade that is used for art fairs and farmers' markets.

In winter, the small event area is flooded and lighted to provide a skating rink, and the information booth beneath the colonnade serves as a warming house and skate rental. The festival park also includes an indoor exhibit and meeting hall. A 930-space parking lot adjoins the festival area. The remainder of the site is a landscaped 16-acre park with parking for 200 cars. The project includes a 17-acre lakefill constructed of sediments dredged from the harbor. The old port breakwaters at the site were enlarged and modified. A pedestrian causeway links the project with downtown.

The $25 million project was financed with private funds (used to finance the marina and festival park), municipal tax increment financing, county revenue bonds, state grants (Wisconsin Department of Natural Resources' Recreational Boating Facilities Fund, Local Park Aid, state community development block grant), and a federal Community Development Block Grant. The marina is owned by the county, but operated by a private firm, Racine Marina Associates, under a long-term license. Converting the harbor's commercial port into a recreational marina required an Act of Congress.

Other developments have built on the success of Racine-on-the-Lake. Gaslight Pointe, under construction near the harbor on the Root River, includes a marina with 120 slips under 99-year leases, restaurants and neighborhood shops, 20 town homes, a 104-unit condominium tower, and a 400-space parking garage. A 150-room hotel is planned for future development.

The developer donated to the city the right-of-way for a path along the riverbank and built a new breakwater to protect against erosion. The $79.5 million project was financed with private funds and municipal tax increment financing, which funded the public improvements at the site. A development agreement between the city and the developer spells out the phasing of public improvements with each element of the project's construction.

Lakeshore Towers is another mixed-use project under construction. It is on the site of a former tank farm on Lake Michigan. The project includes a marina with 52 slips under 99-year leases, a 72-unit condominium tower, and a parking garage. A shoreline promenade along the lakeshore was dedicated to the city by the developer. The site is protected from erosion by a breakwater and revetment. The $17 million development was privately financed.

Racine's most recent Root River development is the Case Corporate Complex, corporate headquarters for J.I. Case, a major farm and construction equipment manufacturer. Its development is occurring in phases.

Phase one is a 120,000-square foot office building and service training center. Phase two will include another office building, a corporate training center, and an exhibition hall, including a visitors' center and museum. Phase three will include an office building and a parking garage. The $100 million development is privately financed.

CONTACTS:

THOMAS WRIGHT, Director of City Development
City of Racine
City Hall
730 Washington Avenue
Racine, Wisconsin 53403
(414) 636-9151

ARNOLD CLEMENT, Planning and Development Director
Racine County Planning and Development Department
14200 Washington Avenue
Sturtevant, Wisconsin 53177
(414) 886-8470

KENOSHA, WISCONSIN

Boating improvements are featured in the waterfront developments of this town, the former headquarters of the American Motors Company. The city has added 200 slips to its Small Boat Harbor on Lake Michigan, reconstructed its existing boat ramp and improved a nine-acre park at the marina. A footpath along the east side of harbor was created, as were two new parking lots. The $830,000 project was financed with municipal general obligation bonds and a grant from Wisconsin's coastal zone management program.

Now under construction is Southport Marina on 60 acres in the port of Kenosha. The marina will have 500-600 slips, a launching hoist, restaurant, and boat repair yard. A 600- to 800-space parking lot as well as dry storage for boats will be provided. Ten acres of the site are being landscaped for a public park, and a mile-long pedestrian footpath will be provided. The site is protected from waves and erosion by two newly constructed breakwaters and a jetty.

The $14 to 16 million project is funded by a local tax increment financing district, a federal Community Development Block Grant and grants from the Wisconsin Department of Natural Resources' Recreational Boating Facilities Fund and the coastal zone management program. Once completed, the city-owned marina will be operated by a private concessionaire. The site was previously a confined disposal facility for harbor dredge spoils.

The city has also completed a 9.4-mile bike trail and footpath, Pike Trail, along Lake Michigan. It links Kenosha's three historic districts (Civic Center, Third Avenue, and Library Park), the municipal marina, and the velodrome and band shell. It was financed by Wisconsin's coastal zone management program and municipal general obligation bonds.

CONTACTS:

RAY FORGIANNI
Department of City Development
625 52nd Street
Kenosha, Wisconsin 53140
(414) 656-8030

DANIEL S. WINKLER, Special Projects Manager
Department of Public Works
625 52nd Street
Kenosha, Wisconsin 53140
(414) 656-8040

MILWAUKEE, WISCONSIN

Milwaukee County is building Lakefront Terrace, a new park which will link the city's downtown district with its lakefront (scale model). MILWAUKEE COUNTY

Milwaukee's waterfront development, long focused on the lakeshore, is expanding to include the renovation of landmark buildings and new construction along the Milwaukee River in the city's downtown.

A key lakefront project now under construction is Lake Terrace, a 7.5-acre park atop a 1,250-space parking structure. The park, to be named the William F. O'Donnell Park, will link downtown Milwaukee with existing city parks along the lakeshore. It will contain landscaped areas, a promenade with vista points overlooking the Lake, a pedestrian overpass leading to the Lake Michigan shore, and an enclosed pavilion housing restaurants, an atrium, and meeting/display space. The $32.5 million project is being funded by private donations from the Greater Milwaukee Committee (a business coalition) and Parks People (a citizen group and a combination of state, County and city funds). The county-owned site is an abandoned railroad right-of-way which was once designated for future freeway development.

McKinley Beach, a 12-acre park on Lake Michigan north of the city's downtown, is also being improved. The project includes a sand beach, a pebble beach designated for windsurfers, landscaping, a playground, a terrace with space for vendors, a vista point overlooking the Lake, and a continuous footpath along the lakefront and into the park. The park is protected from erosion by a variety of shoreline structures, including a revetment, two jetties, and a breakwater. Public access is allowed atop the north jetty. The project, funded by Milwaukee County, is built on 180,000 cubic yards of landfill created with spoils donated by the Milwaukee Metropolitan Sewerage District.

South of the downtown, the new Marcus Amphitheater has been added to Milwaukee's lakefront Henry W. Maier Festival Park. The amphitheater has a covered stage, more than 9,800 seats, and a total capacity of 24,000 people. It serves as the main stage during Summerfest and Milwaukee's ethnic festivals, and hosts other concerts. It is operated by Milwaukee World Festivals Inc. The $13 million amphitheater was financed with municipal revenue bonds.

Development along the Milwaukee River begins at 100 East Wisconsin, where a 37-story office tower rises from the river's banks. It includes a ground-level restaurant and a parking garage. A riverwalk runs along waterfront. The $70 million project was largely privately financed. The city's tax increment financing district contributed a share of the project's riverwalk and a skywalk.

Just upriver is Milwaukee Center, a mixed-use development completed in 1988. It combines the renovated Pabst Theatre with a 6-story, 220-room Wyndham Hotel, shops, a 28-story office tower, the renovation of the historic Oneida Street Power Station into a new theatre, and a parking garage. The separate structures are connected by a glass-enclosed galleria/atrium. Public access along the Milwaukee River is provided by a riverwalk. The $100 million project was privately financed, with help from Milwaukee's tax increment financing district.

Three blocks further along the river is Riverfront Plaza,

which includes the renovation of a former warehouse into offices, a restaurant and specialty shops, as well as a riverwalk and a 30-slip marina with a berth for a dinner cruise tour boat. The $10 million project was financed with a mix of city tax increment financing and private funds.

Schlitz Park, a third mixed-use development on the Milwaukee River, is the 40-acre former Schlitz Brewery complex. Its redevelopment is occurring in phases. Phase one provides 900,000 square feet of offices and a restaurant. The 478,000-square foot Bottle House has been renovated for offices. Later phases will involve the renovation of the Malthouse and Brewhouse.

The site will also include a marina with 114 rental slips, with 40 slips completed in the first phase. A riverwalk and 2,500 parking spaces enhance public access to the river at the site. The $80 million project is financed with private funds, as well as loans backed by municipal general obligation bonds and the city's tax increment financing district.

Milwaukee's downtown system of Riverwalks is a major feature of the city's Milwaukee River renewal. City policy requires that the riverwalks' right-of-way, located on private property along the shoreline, be set aside in new riverfront development. If the footpath is constructed, the developer and the city share the costs up to a maximum city share of 50 percent. The city has riverwalk design guidelines covering construction materials, structural elements, and aesthetics. The 400-foot-long Riverwalk at Marshall Field's and Co. at 101 Wisconsin Avenue typifies these projects. It cost $527,000, or about $1,320 a linear foot.

CONTACTS:

TOM MILLER, Director of City Planning
Department of City Development
P.O. Box 324
Milwaukee, Wisconsin 53201
(414) 223-5900

BRIGID SULLIVAN, Director
Milwaukee County Parks Department
9480 Watertown Plank Road
901 N. 9th St.
Wauwatosa, Wisconsin 53226
(414) 257-4501

ELIZABETH BLACK
Milwaukee World Festivals Inc.
200 N. Harbor Drive
Milwaukee, Wisconsin 53202
(414) 273-2680

SHEBOYGAN, WISCONSIN

Development in Sheboygan showcases the city's commercial and sportfishing heritage. At the city's Inner Harbor, where the Sheboygan River flows into Lake Michigan, planners have proposed a new Sheboygan Marina and DeLand Park festival site. It will include a 438-slip municipal marina and redevelopment of DeLand Park into a festival site, including construction of waterfront footpaths.

Development of the marina has been delayed because sediments in the Inner Harbor are contaminated with PCBs.

Shanty Village, a mixed-use project located further up the Sheboygan River, is the city's old commercial fishing district. The village features eleven commercial fisherman's shanties which have been restored or newly constructed for specialty shops, restaurants, and office space. A 76-slip city-operated marina, including three slips for commercial fishermen, is located along the riverfront. Net drying racks and work areas are also provided for commercial fishermen. The 31-room Harbor Inn adjoins the shanty development. A 2,600-foot boardwalk along the river's edge connects the retail area to the two-acre riverside Rotary Park.

The $12 million development was financed with private funds, the city's tax increment financing district, a grant from the Wisconsin coastal zone management program, a loan from the U.S. Economic Development Administration, and a federal Community Development Block Grant.

Other riverfront developments in Sheboygan include West Bay Trading Company, a furniture factory and warehouse rehabilitated for shops and offices, and Riverfront, a mixed-use project with shops, a restaurant and offices in four buildings. The city's Riverwalk runs along the Sheboygan River through the Riverfront project. The city used federal Community Development Block Grants to construct a 120-space parking lot adjacent to West Bay Trading Company and to purchase and clear the Riverfront site. The construction of both projects was privately financed.

CONTACT:

ROBERT PETERSON, Director of City Development
City of Sheboygan
828 Center Avenue
Sheboygan, Wisconsin 53081
(414) 459-3377

MANITOWOC, WISCONSIN

Manitowoc has continued to improve a waterfront that is among the most attractive and successful on the Great Lakes. The Manitowoc Municipal Marina expansion is adding 50 slips and 350 parking spaces to the city's existing 200-slip marina on Lake Michigan. An additional 800 feet of bulkhead are being installed as part of the project. The expansion will allow the addition of up to 200 more slips to the marina as demand warrants. The $1 million marina expansion was financed from the city's general fund and a grant from the Wisconsin Department of Natural Resources' Recreational Boating Facilities Fund.

The city has also extended its 1.75 mile-long Manitowoc Walkway System along the Manitowoc River, Little Manitowoc River and Lake Michigan. The walkway system is 95 percent complete, and connects with walkways at two existing projects, the Manitowoc Maritime Museum and the Inn at Maritime Bay.

The riverfront walkways were completed in 1987, and the lakefront walkways in 1989. Building the walkways cost $250,000, with funds coming from Manitowoc's general fund, Wisconsin's coastal zone management program, a federal Community Development Block Grant, and the Small Business Administration. Unemployed workers hired through a Small Business Administration employment program installed landscaping along the paths.

CONTACT:

NICHOLAS LEVENDUSKY, Deputy City Planner
Department of Planning
P.O. Box 1597
Manitowoc, Wisconsin 54221-1597
(414) 683-4435

TWO RIVERS, WISCONSIN

Redevelopment of the Two Rivers waterfront is beginning with projects that improve recreational opportunities. The city has undertaken three projects: Harbor Dock, a recreational fishing dock in Two Rivers Harbor; the Vets Park Launch Facility, a four-lane boat ramp and 1,200-foot footpath on the West Twin River; and the 1,000-foot West Twin River walkway.

Funds for the projects came from Two Rivers' general fund, the Wisconsin Development Fund, the State Department of Natural Resources' Recreational Boating Facilities Fund and coastal zone management program, and the federal Dingle-Johnson program. The right-of-way for the West Twin River walkway was donated to the city by Egger Industries, a local manufacturer.

CONTACTS:

STEVEN T. NENONEN, City Manager
P.O. Box 87
Two Rivers, Wisconsin 54241
(414) 793-5532

DAN PAWLITZKE, Economic Development Supervisor
P.O. Box 87
Two Rivers, Wisconsin 54241
(414) 793-5565

STURGEON BAY, WISCONSIN

Sturgeon Bay, long a center of shipbuilding on the Great Lakes, is renovating its Westside Waterfront Parks—including Sawyer Park, Bayview Park, and the West-Side Dock—on Sturgeon Bay. They are being improved in a coordinated program that will provide a festival facility, museum site, six-lane boat launch with 250 parking spaces, modification and addition of bulkheads to accommodate recreational boating use of the municipal dock, a two-story Wisconsin Department of Natural Resources fisheries building, and a dockmaster building. A handicapped walkway and fishing pier were constructed in Bayview Park.

The $2.1 million project is being financed with municipal general obligation bonds and Department of Natural Resources grants.

CONTACT:

NORB SCHACHTER, Mayor
City of Sturgeon Bay
30 South Third Avenue
Sturgeon Bay, Wisconsin 54235
(414) 743-5984

GREEN BAY, WISCONSIN

The Fox River, the principal tributary to Green Bay, is the site for this city's waterfront revitalization efforts.

More than a mile of riverwalks has been constructed along the Fox at Admiral Flatley Park and Veteran's Park. Their $130,000 cost was financed with municipal general obligation bonds, tax increment financing, and a grant from Wisconsin's coastal zone management program. The city's downtown waterfront plan calls for the construction of additional riverwalk segments as the waterfront redevelops.

The riverwalk has spurred additional development on the shoreline, including Riverwalk Plaza, an office development. Riverwalk Plaza's first phase, a 36,000-square foot office building and 125-space parking lot, has already been built. The second phase will be a 21,000-square foot office building. The riverwalk right-of-way through the project site was donated to the city by the developer. The project's $3 million first phase is privately funded.

The Fox River Docks, a new municipal marina adjoining Admiral Flatley Park, has 52 slips. As part of the marina improvements, the city added 3.5 acres to Admiral Flatley Park, including a lighted fountain, benches, and another 800 feet of riverwalk. The $266,000 marina was financed with municipal general obligation bonds. It is managed by the Holiday Inn adjoining the marina.

The city has also improved its boat launching ramp on the Fox River. The project includes the addition of three launching ramps and 200 parking places. The site is owned by the Green Bay Metropolitan Sewerage District.

CONTACTS:

DALE PRESTON, Principal Planner
City of Green Bay
Planning Department
100 N. Jefferson Street
Green Bay, Wisconsin 54301
(414) 448-3400

ALAN P. JOHNSON, Port Director
Brown County Port Authority
305 E. Walnut Street
Green Bay, Wisconsin 54301
(414) 448-4290

MARINETTE, WISCONSIN

A new boat launch and one-acre park have been added at the Sixth Street Slip on the Menominee River, a tributary to Green Bay. The city-owned site had previously been used as a dump. An existing launching ramp at Boom Landing has also been improved.

The projects were funded by Marinette's general fund, in-kind municipal contributions provided by the use of city-owned land, Wisconsin's coastal zone management program, federal Dingle-Johnson funds, a contribution from the Menominee and Marinette Sport Fishing Association, and a donation-in-kind from a local contractor.

CONTACT:

GEORGE COWELL, P.E., City Engineer and Planner
1905 Hall Avenue
Marinette, Wisconsin 54143
(715) 735-7427

MENOMINEE, MICHIGAN

Menominee, just across the river from Marinette, has improved its Municipal Marina on Green Bay. Forty slips were added to the existing 80-slip marina, and the marina's 5,700-foot-long breakwater was rebuilt. Additional parking is planned.

The $2.65 million project was funded by the marina concessionaire, the city's downtown tax increment financing district, private donations from local corporations and individuals, and grants from the Michigan Department of Natural Resources' Waterways Fund and Natural Resources Trust Fund. The concessionaire funded the slip additions and the municipal and state funds paid for the breakwater repairs.

CONTACTS:

DON SANDSTROM, Planning Director
City Hall
2511 10th Street
Menominee, Michigan 49858
(906) 863-2656

JOHN JOHNOWITZ, City Engineer
City Hall
2511 10th Street
Menominee Michigan 49858
(906) 863-2656

ESCANABA, MICHIGAN

Escanaba has redeveloped the marina at its Municipal Harbor on Little Bay de Noc at the north end of Green Bay. The project involved dredging the harbor and constructing finger docks with 150 rental slips. The $760,000 project is being financed by the city and the state's Waterways Commission and Natural Resources Trust Fund. Dredge spoils from the harbor were used to replenish the municipal beach.

CONTACT:

JOHN DULEK, City Engineer
City of Escanaba
City Hall
Escanaba, Michigan 49829
(906) 786-9402

CHARLEVOIX, MICHIGAN

This popular resort has improved its Ferry Avenue Park on Lake Charlevoix on the northeast shore of Lake Michigan. The park was expanded onto a newly acquired parcel, the launching ramp was renovated, and the park was rehabilitated. Future developments include a 1,500-foot extension of a lakefront footpath, and a new 160-space parking lot.

The $1 million project was financed by the Michigan's Natural Resources Trust Fund and the city. The city was required to protect wetlands within the site as a condition of its permits from the Michigan Department of Natural Resources and the U.S. Army Corps of Engineers.

CONTACT:

MIKE WISNER, City Manager
City Hall
210 State Street
Charlevoix, Michigan 49720
(616) 547-3270

TRAVERSE CITY, MICHIGAN

The West Beach Project, on 5.2 acres on Lake Michigan's West Grand Traverse Bay, is one of the few urban waterfront projects in the basin emphasizing restoration of a natural landform, rather than development, dredging, or landfilling. The project restores a half-mile-long beach from a former railroad right-of-way, sculpting berms to replicate low sand dunes, and replanting the site with native vegetation. A new 30-space parking lot to support the beach was built. Half-mile-long bicycle and pedestrian paths constructed along the beach several years ago will connect the park to 13 miles of other bike paths on the city's waterfront and a state-owned former railroad right-of-way.

The $374,000 project was funded by a grant from the Michigan Natural Resources Trust Fund and the city, which provided labor and materials.

Traverse City's other waterfront improvement is the 300-foot Boardman Riverwalk along Boardman River, a tributary to Grand Traverse Bay. The boardwalk, located on a 15-foot-wide pedestrian easement and 66-foot-wide right-of-way, crosses behind buildings backing on the river and runs under the state highway bridge near mouth of the river. The riverwalk links a city beach on Grand Traverse Bay to the Boardman neighborhood, providing access from several bayfront hotels to the downtown. It is handicapped-accessible and will be lit at night.

The $73,000 project was financed by the city's Downtown Development Authority, a grant from the Michigan Department of Natural Resources' coastal zone management program, the contribution of services and materials from the city, and donations from Traverse City Light and Power and the Dr. James Whitney Hall Fund, a local environmental foundation. A local business donated the pedestrian easement needed for the project.

CONTACTS:
RUSSELL SOYRING, Director of Planning
400 Boardman Avenue
P.O. Box 592
Traverse City Michigan 49685
(616) 922-4465

CHARLES JUDSON, Chairman
Downtown Development Authority
P.O. Box 2337
Traverse City, Michigan 49685
(616) 922-4446

MANISTEE, MICHIGAN

Manistee is undergoing one of the most substantial waterfront revitalizations on Lake Michigan. The city has redeveloped the south shore of the Manistee River, adjoining its central business district. Now development is expanding to the river's north bank and the Lake Michigan shoreline. Manistee's waterfront development financial strategies are among the most diversified of any small city in the basin.

The Southside Riverwalk, now under construction, will stretch 1,400 feet along the Manistee River, connecting the city's central business district with Lake Michigan. The right-of-way was previously occupied by the back lots of commercial buildings. The riverwalk's improvements include a boardwalk, Victorian street lighting, and a bulkhead to protect against erosion. Eight new boat slips have already been installed along the bulkhead.

The $800,000 project was financed by municipal revenue bonds to be repaid from renegotiated leases on existing boat slips along the walkway, a Community Development Block Grant, Downtown Development Authority bonds to be repaid by a tax increment financing district, street funds, in-kind donations from the city, a loan from the city's marina fund to be repaid by rental of the eight slips, and a grant from the Michigan Department of Natural Resources recreation bond program. The city used coastal zone management funds to plan future phases of the project.

Other new developments along the river include Johnston's Marina with 27 slips (including three charter boats). A bubbler protects the docks from winter ice. Fluctuating lake levels were not anticipated in the project's design, requiring additional dredging when lake levels became lower.

The Harbor Village at Manistee Beach is a mixed-use marina and residential project on 55 acres of city land on Lake Michigan and the Manistee River. Project plans call for a 380-slip dockominium marina opening on the Manistee River, 140 condominiums, 35 single family homes, and a 100-room time-share motel. A 2,700-foot public accessway will run along the waterfront throughout the project.

The development's architecture will emphasize Victorian themes compatible with Manistee's downtown. The project was planned jointly by the city and a developer. The $44 million project is expected to be financed largely by the developer, with the city contributing infrastructure improvements. Other city costs are expected to be repaid through the sale of the property which was previously used for sand mining.

The Northside Riverwalk is a three-acre public park on the Manistee River, upriver of the Harbor Village site. The project included a 1,500-foot riverwalk, a gazebo bandstand, a 120-car parking lot, and landscaping. The $140,000 project, part of the city's central business district plan, was financed from city's general fund and private contributions, portions of which were raised by the Jaycees. Much of the labor used to develop the park was also donated.

The Manistee Inn and Marina, a 26-unit motel with shops and a 15-slip marina, is located on the Northside Riverwalk. The project includes a 50-space, two-story parking lot used for boat storage in the winter. Most of the $1 million development was privately financed, while the city's tax increment financing district funded part of the parking structure.

An 18-unit condominium project is also under construction on the river's north bank. The first six-unit phase has been completed, while work on the remaining 12 units has not yet begun.

Manistee's other principal waterfront development is the 17-lot Dunes Subdivision on 25 acres of undeveloped city land on Lake Michigan. The city will retain a 12-acre, half-mile-long lakefront beach, with public access provided through numerous access points. Because the site is in a high-risk erosion zone, the building lots are set back from the beach.

The city purchased the property in 1972 as part of a larger site, a portion of which has been developed as an industrial park. The original property acquisition cost $250,000; since then, some of the industrial property was sold for $250,000 and the Dunes Subdivision site is being sold for $500,000. In addition, revenues from natural gas wells drilled on the property have been deposited in a $3.6 million city trust fund.

CONTACTS:

BEN BIFOSS, City Manager
70 Maple Street
Manistee City Hall
Manistee, Michigan 49660
(616) 723-2558

CHARLES ECKENSTAHLER
Abonmarche Group
95 West Main Street
P.O. Box 1088
Benton Harbor, Michigan 49022
(616) 927-2295

MUSKEGON, MICHIGAN

Four projects are now under construction on Muskegon Lake's waterfront, and two more are planned. Among the largest is Terrace Point, a mixed-use project on 35 acres. The project's first phase, already completed, includes a 114-slip marina, a restaurant, and the headquarters office of the SPX Corporation. A boardwalk provides public access along the waterfront. Future phases are planned to include duplex condominiums. The project's $22 million first phase was privately financed.

Bluffton Bay is another mixed-use project under construction. Its first phase includes a 120-slip marina and renovation of an existing restaurant, while 48 apartments are planned for the second phase. Nearby is the 32-acre Great Lakes Marina. Its first phase, already completed, has 125 slips and a boat sales yard. A second phase of 212 slips is planned. At Lakeshore Harbour, 12 condominiums have been built next to the existing 63-slip marina.

Development of an additional 120 acres of Muskegon Lake's shoreline is now being planned. Surplus land near the Consumers Power Company's B. C. Cobb power plant will be developed into Shoreside Harbor, a 380-slip marina with 200 condominiums. The project is expected to cost $40 million. Harbour Towne, planned for 66 undeveloped acres, will include a 128-slip marina and 220 condominiums. The city intends to provide $1.4 million in street and utility improvements, financed with special assessment bonds, at the site.

CONTACTS:

RICK CHAPLA, Director
Department of Planning and Community Development
City of Muskegon
933 Terrace Street
Muskegon, Michigan 49443
(616) 724-6702

CHARLES ECKENSTAHLER
Abonmarche Group
95 West Main Street
P.O. Box 1088
Benton Harbor, Michigan 49022
(616) 927-2295

GRAND HAVEN, MICHIGAN

This city, known for its sunny beaches, is continuing one of the most successful waterfront revitalizations on Lake Michigan. Its most recent development is Harbourfront Place, an old piano factory on the Grand River which has been rehabilitated into shops, offices, and 80 condominiums. The city installed 125 parking spaces in two lots adjacent to the project. Most of the $13.5 million project is privately financed. Public financing included an Urban Development Action Grant and Small Cities Block Grant, as well as city financing of the parking lots adjoining the project.

CONTACT:

JERRY LIETZKE, Director of Planning
City of Grand Haven
414 Washington Street
Grand Haven, Michigan 49417
(616) 842-3210

ST. JOSEPH, MICHIGAN

A new marina has been built on the St. Joseph's River, the city's Lake Michigan tributary, and a new boat ramp is planned. Pier 33, a private marina, has 222 boat slips, including 100 transient slips, inside boat storage, and 400 parking spaces. The storage area provides winter and summer boat storage and repair, as well as retail boat sales. The $8 million development was privately financed. The site is an abandoned railroad right-of-way.

The boat ramp, planned to accommodate 124 cars and trailers, is expected to cost $324,000. Its cost will be shared by the city and the Department of Natural Resources Waterways Fund.

CONTACT:

NORM ORBECK, Economic Development Director
City of St. Joseph
620 Broad Street
St. Joseph, Michigan 49085
(616) 983-6324

BENTON HARBOR, MICHIGAN

No development reported.

NEW BUFFALO, MICHIGAN

The South Cove "dockominiums" in New Buffalo allow boatowners to step from their living rooms onto their decks, a concept adopted by a number of Lakes communities. HARRY WEESE & ASSOCIATES

At New Buffalo, the construction of harbor jetties at the mouth of the Galien River and the improvement of the road along the river has unleashed marina and vacation condominium development. Unlike most other Lake Michigan cities, most projects at New Buffalo are occurring on previously undeveloped lands, including some wetlands and dunes. The city's boom began in the early 1980s, driven by Chicago-area residents seeking weekend retreats.

At the 18-acre city park at the Galien River's outlet into Lake Michigan, the beach and boat launching ramp have been improved. A new concession stand, restrooms, 250-car parking lot, and a footpath and wooden stairway through the adjoining dunes have been added. The launching ramp, located on the riverfront just inland of the shoreline road, was reconstructed to provide eight launching lanes and parking for 300 boat trailers. The $522,000 project was financed by the city, the Michigan Department of Natural Resources' Division of Waterways and coastal zone management programs, and private donations. A transient marina with 20 moorings is also planned for the site.

Four projects with more than 600 marina slips, mostly dockominiums, and 480 condominiums, are under construction or have recently been completed along the river. At South Cove, 106 residences and 47 slips of a planned 400-unit residential/marina project have been completed. The residences include a mix of condominiums and small garage/cabana units. Dunewood, another residential/marina project under construction on a private channel of the Galien River, will have 58 town homes and condominiums and a 44-slip rental marina. The riverfront Harbor Point Condominiums have three ten-unit buildings and a 60-slip dockominium marina. The nearby Harbor Landing marina has 100 dockominium slips.

A variety of regulations control Galien River marina development. The city has a special marina zoning designation, and the Michigan Department of Natural Resources has conditioned its permits to require that the marinas be dredged into the riverbank, so that the slips do not extend into the river. Developers have also been required to dedicate easements over the shorelines opposite their projects so that slips could not be developed on both sides of the river.

Second-home development is also occurring in the dunes along New Buffalo's Lake Michigan coast. At Warwick Shores, 74 of 162 planned residences have been built on a 40-acre site. A submerged offshore stabilizer, tied to a revetment, was built to protect against erosion.

CONTACT:
THOMAS JOHNSON, City Manager
224 W. Buffalo Street
New Buffalo, Michigan, 49117
(616) 469-1500

MICHIGAN CITY, INDIANA

In Michigan City, development is occurring both along the Lake Michigan shoreline and through redevelopment in the city's harbor. There are two condominium developments on the lakefront: Sunset Point with 14 units (eight constructed; 6 planned), and Dunescape, an 11-story tower with 68 condominiums and a parking garage. The projects represent a $12 million private investment. A 250- to 300-foot shoreline setback protects the developments from erosion. At both projects, land disturbed during construction has been replanted with dune grass. Public access is allowed on the beach in front of Sunset Point, where the developer granted an easement allowing the city to operate beach cleaning equipment.

Indiana's Department of Natural Resources has built a new Division of Fish and Wildlife Building in the Michigan City harbor. The site was previously occupied by a grain elevator. The building contains offices, research laboratories, living quarters for resident researchers, and moorings for research vessels and law enforcement patrol boats. Public access for fishing is allowed atop parts of the 300-foot-long seawall which protects the development against erosion. Construction of the $1.8 million project was financed from the state's general fund and the federal Sport Fish Restoration Act (Wallop-Breaux) funds. The site was a gift from a private citizen to Michigan City, which subsequently donated the property to the state.

CONTACTS:
JOHN PUGH, City Planner
Department of Planning
City of Michigan City
100 East Michigan Boulevard
Michigan City, Indiana 46360
(219) 873-1419

DON BRAZO, Fisheries Biologist
Department of Natural Resources
100 West Water Street
Michigan City, Indiana 46360
(219) 874-6824

GARY, INDIANA

Gary is proposing to construct a mixed-use Tourist Center on 400 acres of idle USX Corporation property on Lake Michigan.

Because the top priority is to create boating facilities, the development's first phase, begun in April 1991, will include a launch ramp, fishing pier, access road, and utility improvements. Other project plans are still being developed. A 2,500-slip marina, bathing beach, pedestrian paths, boat sales, a restaurant, specialty shops, apartment buildings, a 250-room hotel, and golf course have been suggested for later phases of the project.

Gambling casinos could be located here if approved by the Indiana legislature. A marina service facility with a boat repair yard and dry storage for up to 450 boats, festival market, aquarium, and an 1,800-seat outdoor amphitheater have also been proposed for the site. A new breakwater would be constructed to protect the area from wave attack and erosion. A landscaped berm, constructed of slag and lime residue, is planned to separate the development from the USX steel mill.

Gary and USX have signed a pre-lease agreement and are now negotiating provisions for the first phase of the project. An environmental study on the development has been prepared by the National Park Service, which operates the adjoining Indiana Dunes National Lakeshore. The total project is expected to cost $150 to $160 million.

CONTACTS:

ARLENE COLVIN, Director
Division of Physical and Economic Development
City of Gary
401 Broadway
Gary, Indiana 46402
(219) 881-1425

BARBARA WAXMAN
Lake Michigan Marina Development Commission
8149 Kennedy Avenue
Highland, Indiana 46322
(219) 923-1060

EAST CHICAGO, INDIANA

East Chicago has a brand new small boat harbor, the Robert E. Pastrick Marina, on Lake Michigan. It includes 286 rental slips and 20 transient slips, three launching ramps, a dry storage facility, marina offices, and a marina store. A 307-space parking lot, with 42 free spaces and the remainder paid parking, is provided. Public fishing access is provided atop the breakwater. Stairs and ramps have been provided to make the breakwater's walkway fully accessible to the handicapped. Comfort stations, fish-cleaning stations, a restaurant, a boat well and a travel lift will be constructed in 1991.

The Old Fishing Pier adjoining the marina was also renovated. New flooring and lighting were added and the pier was made handicapped accessible.

The $14.5 million project was financed with general obligation bonds issued by the city's Park Department and Redevelopment Commission. The Lake Michigan Marina Development Commission, an Indiana state agency, provided $190,000 for project design and $1.5 million for amenities and improvements.

CONTACTS:

NICK TRGOVICH, Marina Director
3301 Aldis
East Chicago, Indiana 46312
(219) 391-8482

BARBARA WAXMAN
Lake Michigan Marina Development Commission
8149 Kennedy Avenue
Highland, Indiana 46322
(219) 923-1060

HAMMOND, INDIANA

The recently completed Hammond Marina is among the largest on Lake Michigan, with 1,113 rental slips, a five-lane boat ramp, and two service buildings. Tour and party boats can be moored at the marina's head piers. The *Milwaukee Clipper*, a Lake Michigan carferry listed on the National Register of Historic Places, has been relocated to the marina and will be renovated to provide office and retail space.

An innovative 3,000-foot-long tandem breakwater with a submerged outer "reef" breakwater is being constructed to protect the boat basin from wave attack. Fishermen can use a handicapped-accessible walkway with three fishing platforms atop the breakwater.

The marina was planned and developed by the city's port authority. Its $23 million cost is being funded with private lease-back financing, a loan backed by general obligation bonds from the city's redevelopment commission, a grant from the state's Lake Michigan Marina Development Commission, a loan from the state general fund, and a grant of federal Dingel-Johnson funds. The Northern Indiana Public Service Company provided a bridge loan.

CONTACTS:

LEWIS TIMBERLAKE
Hammond Port Authority
649 Conkey Street
Hammond, Indiana 46320
(219) 933-7678

BARBARA WAXMAN
Lake Michigan Marina Development Commission
8149 Kennedy Avenue
Highland, Indiana 46322
(219) 923-1060

CHICAGO, ILLINOIS

Chicago leads the region in waterfront development, with a wave of new construction focusing on the Chicago River. THE LEVY ORGANIZATION

A major upgrading of Chicago's waterfront is underway. Recent developments, as well as projects presently proposed, focus on the Lake Michigan shore as well as the banks of the Chicago River.

The renovation of historic Navy Pier highlights lakefront development. The pier's bulkheads and its East End and Head House structures are now being renovated. Other cargo sheds have been demolished. New development is planned to include an exposition and festival hall, a wintergarden and arcade enclosing a year-round park, and specialty shops and restaurants, flexible indoor space for museum and cultural facilities, and a theatre. A marina, including berths for the seven tour boats now moored at the pier, is also being considered. The $150 million project is financed by a grant from the state's Build Illinois Fund.

Another lakefront development is the expansion of the John G. Shedd Aquarium. The project adds an "Oceanaria" including whales and other marine life. A walkway will provide public access along the shore. The expansion includes 1.8 acres of landfill. The $43 million project is funded by private contributions, the State of Illinois, and the Chicago Park District.

Improvements to lakefront boating facilities have been undertaken by the Chicago Park District, which operates the lakefront parks. At Burnham Harbor, south of the Shedd Aquarium, there is a new fish cleaning station, harbormaster's office, and concession facility. Further south, the park district is renovating the Jackson Park Coast Guard Station to house a marine fueling station, concession space and office. The two projects' $1.3 million cost is being financed by the park district and the Illinois Department of Conservation.

At the mouth of the Chicago River the park district has proposed the Chicago River Turning Basin marina, a marina and park. Present plans call for construction of a 500-slip Monroe Marina south of the river, followed by a new marina, Chicago Moorings, on the river's north side adjoining Navy Pier. The district expects to fund the $12 million Monroe Marina with revenue bonds.

Still being developed are park district/U.S. Army Corps of Engineers plans for renovation of the seawalls and other structures protecting the lakefront from shoreline erosion.

Cityfront Center, a $2 billion, 60-acre mixed-use development just inland of the Chicago River turning basin, is the largest waterfront project in the region. It includes five major components: NBC Tower, a 40-story office tower completed in 1990; an additional office tower, still in the early design stage; the Sheraton Chicago Hotel and Towers, a 1,200-room hotel under construction; Cityfront Place, three mid-rise and high-rise buildings holding 900 apartments also under construction; and North Pier, two separate structures including North Pier Chicago, an old waterfront warehouse rehabilitated into specialty shops and restaurants, and the 60-story, 505-unit North Pier Apartments.

Transient moorings for tour boats and pleasure craft are provided along the navigation channel of Ogden Slip, which abuts North Pier Chicago. Five acres of parks, including Ogden Plaza and DuSable Park, as well as a milelong waterfront promenade, provide pubic access along the river and the Ogden Slip. A fountain depicting the watershed of the Great Lakes and Mississippi River is located at a riverfront plaza along the promenade. North Pier Chicago also holds the Children's Museum and Chicago Maritime Museum. A waterfront drive and connecting roads were extended through the site as part of the project.

Cityfront Center is privately financed. The Metropolitan Water District of Chicago contributed $2.5 million of the $3 million cost of the fountain. While most of the parks and the waterfront promenade were donated to the Chicago Park District, the Cityfront Center Maintenance Association will continue to maintain them using dues assessed against the projects' lessees.

Another recently completed riverfront development, the Illinois World Trade Center, includes a 350-room Hotel Nikko, and the high-rise offices of Quaker Tower. Future phases may include high-rise apartments and another office tower. The project's riverfront promenade includes several small plazas, outdoor dining areas, and a small retail area.

Wolf Point, at the confluence of the Chicago River's north and south branches, is the planned site of Riverbend Complex, which will include a corporate campus, hotel and retail space. The buildings will center on a one-acre riverside plaza that steps down to a shoreline promenade. Slips for pleasure boats will be constructed if they do not interfere with commercial navigation on the river.

The Morton International Building is a recently completed office tower on the Chicago River's south branch. The project includes offices, shops, a restaurant, and a parking garage. Public access along the river is provided by a riverfront promenade and a small city park, which has been rehabilitated with benches, trees, and lighting, and will be maintained by the developer. At street level, a covered arcade and an outdoor plaza with lookouts onto the park and river run the length of the site. The privately financed, $80 million development was constructed on air rights over railroad right-of-ways.

The Chicago River Cottages, on the river's north branch, provide a small scale contrast to the skyscrapers being built along the rest of the river. The cottages are a four-unit, single-family residential development.

CONTACTS:

CAROLYN JOHNSON/HAL HARVEY
Metropolitan Pier and Exposition Authority
2301 South Lake Shore Drive
Chicago, Illinois 60616
(312) 721-6150

Chicago Park District
425 East McFetridge
Chicago, Illinois 60605
(312) 294-2263

JIM DEMANCHIK, Riverfront Planner
City of Chicago
City Hall, Planning and Development
121 N. LaSalle Street, Rm 1003
Chicago, Illinois 60602
(312) 744-4453

AMY HECKER, Vice-President for Operations
Cityfront Center
Chicago Dock and Canal Trust
401 N. Michigan Avenue, Suite 3145
Chicago, Illinois 60611

EVANSTON, ILLINOIS

No new development reported.

WILMETTE, ILLINOIS

No new development reported.

HIGHLAND PARK, ILLINOIS

A 694-acre U.S. Army base, Fort Sheridan, on the Lake Michigan shore of Highland Park and Lake Forest, will be closed in 1994. Mixed-use redevelopment of the fort is recommended by both the Fort Sheridan Commission, an advisory committee of local officials, and a local non-profit organization, The Advocates for the Public Interest in Fort Sheridan. A variety of specific proposals, including protection and rehabilitation of The Fort's historic structures for research facilities and administrative uses, and use of the fort's open space for a park, golf course, or military cemetery are still being debated.

The fate of the 496 single- and multi-family residential units on the base must also be decided. There is general consensus that the 1.75-mile shoreline on the base should be opened to public access, and that 100 acres of ravines and bluffs should be protected as a natural habitat. The federal Department of Defense wants to retain 90 acres for military training.

The Advocates have proposed that the site be redeveloped and operated by a special purpose agency representing local governments. The group recommended that the $59 million redevelopment of the site be financed with revenue bonds.

CONTACT:

DEBORAH PEARSON
Planning Department
City of Highland Park
505 Laurel Avenue
Highland Park, Illinois 60035
(708) 432-0800

NORTH CHICAGO, ILLINOIS

Harbor Club is a proposed residential/marina development on 76 acres of former industrial land on Lake Michigan. The project would include 330 town homes, each with its own boat slip. A clubhouse and bathing beach would also be developed. A new breakwater would be installed to protect the site from wave attack and erosion. The $150-$200 million project would be privately financed.

CONTACT:

ROBERT STONE, Building Commissioner
City of North Chicago
1850 Lewis
North Chicago, Illinois 60064
(312) 578-7778

WAUKEGAN, ILLINOIS

No new development reported.

LAKE HURON
Including the St. Clair and Detroit Rivers

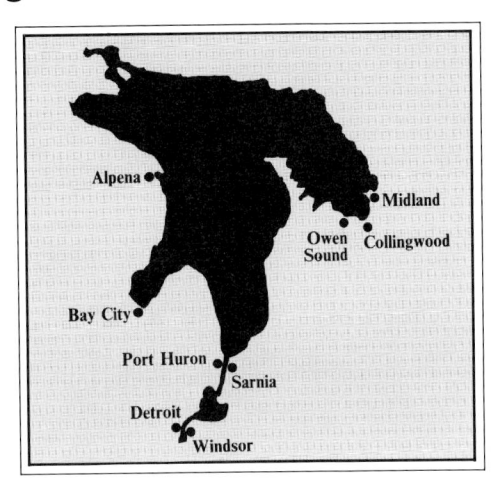

Two kinds of development predominate on Lake Huron. In Georgian Bay, and to a lesser extent on the northeast coast of Michigan, are marinas and second-home developments catering to summer-time visitors from Toronto and other metropolitan areas. Along the busy St. Clair River are condominiums, parks, and cultural facilities that provide a window on the water for the manufacturing cities of Sarnia, Port Huron, Windsor, and Detroit.

More than 935 acres (379 hectares) of waterfront have been renovated or built upon along Lake Huron and the St. Clair River since 1986, for a total of more than $61.2 million ($C70 million) in new construction. That includes more than 2,000 new condominiums and apartments, 1,650 new boat slips, 113,000 square feet (31,500 square metres) of shops and restaurants, and 150 hotel rooms. Seventeen acres (seven hectares) of new or rehabilitated parks, and more than 2.5 miles (four kilometres) of new waterfront pathways have been added.

Better things are yet to come. Plan for future construction call for more than $200 million ($C230 million) in waterfront projects on 96 acres, including another 2,000 condominiums and apartments, 950 boat slips, 307,000 square feet (85,600 square metres) of shops, several new hotels, and six more miles (10 kilometres) of waterfront walkways.

Lake Huron's most noteworthy developments include:

- Detroit's Linked Riverfront Parks, where arts and community events enhance public enjoyment of the waterfront.
- Midland and Collingwood, on Georgian Bay, where almost 2,600 condominiums and more than 600 new boat slips for weekending Torontonians have been built or proposed.
- Sarnia, which is proposing a combined center for environmental interpretation and the performing arts.

DETROIT, MICHIGAN

Detroit's new Linked Riverfront Park System provides three parks totalling 32 acres on the Detroit River. The three parks—St. Aubin Park, Chene Park, and Mt. Elliot Park—are being linked by a mile and a quarter of bicycle paths on existing surface streets.

Twelve-acre St. Aubin Park, the closest to downtown, was developed from 1987-89. It contains a 67-slip marina for transient boaters and six acres of landscaped open space with picnic shelters.

At Chene Park, the outdoor amphitheater was expanded in 1988-1990 to hold 7,000 people. A berm on the park's western edge contains dredge spoils from riverfront construction. A cement plant storage silo separates Chene Park from St. Aubin Park.

Mt. Elliot Park is located one-half mile further upriver next to a United States Coast Guard station. Through a land swap with the Coast Guard the city added to the park's acreage and received several buildings that will become a park interpretive and visitor center. Park construction will begin in 1991.

Financing for the project was complex, drawing on a variety of federal, state, and local revenue sources. Construction costs at Chene Park alone were $12 million.

Porterfield Marina Village is a mixed-use project under construction on city land on the Detroit River. The site, currently occupied by a city park with a small marina, is being redeveloped to hold a new 450-slip marina, a Shooter's restaurant, and a yacht club. An agreement with the Michigan Department of Natural Resources reserves 350 slips for public use. Englereed Park and a boat ramp at the site will be retained. Future phases of the project may include 220 residences (100 apartments and 120 town house condominiums) and a shopping center.

The bulk of the $33 million project is being financed privately. Detroit has loaned $1.9 million from a UDAG grant to the developers, with whom it has signed a 50-year lease for the parcel. All but 18 of the 180 new jobs expected to be created by the project are CETA eligible, and 90 of them are reserved for minority employees.

Harbortown is another big mixed-use project on the Detroit River. Its completed first phase included two apartment towers with 280 residences, 200 town houses, and shops. The second phase, a 150-slip marina, is now under construction. Future phases may include up to 700 additional residences.

CONTACTS:

RICHARD RYBINSKI, Urban Development Assistant
Department of Community and Economic development
150 Michigan Avenue
Detroit, Michigan 48226
(313) 224-2304

CLARENCE LEE, Planner
Department of City Planning
2300 Cadillac Tower
Detroit, Michigan 48226
(313) 224-6380

JOHN JONES, Planning and Grants Unit
Recreation Department
735 Randolph Street, Rm 1702
Detroit, Michigan 48226
(313) 224-1122

PORT HURON, MICHIGAN

Port Huron's Cross Point Condominiums are being built on the site of former warehouses. CITY OF PORT HURON

In Port Huron, the Thomas Edison Parkway and Inn has been built on 42 acres of city land on the St. Clair River. The 7-acre parkway has a scenic drive and a riverfront path with fishing access and vista points for viewing passing ships. Footpaths will eventually connect it with Pine Park to the south, to make a 1.5-mile walkway along the river. The Thomas Edison Inn is a 150-room hotel, restaurant, and meeting facility. Twenty acres of the site, formerly a cement factory and railroad right-of-way, remain undeveloped. The city sold the hotel site to the developer, who financed the hotel privately.

Development is also occurring along the Black River, a tributary to the St. Clair River. Projects there include Cross Pointe Condominiums, 31 town houses with their own boat slips. Additional units will be built in the future. The site, once occupied by residences and warehouses, was acquired by the city and sold to the builder for redevelopment.

Also on the Black River is the Port Huron Municipal Marina, where 120 slips have been added to the existing 254-slip harbor. The $1.1 million marina was funded by a city loan and a grant from the Michigan Waterways Commission.

CONTACT:

RAYMOND STRAFFON, Director
Planning Department
City of Port Huron
100 MacMorran Boulevard
Port Huron, Mich 48061
(313) 987-6000

BAY CITY, MICHIGAN

Bay City's Municipal Pier, built on the site of an old railroad bridge, provides a vista point and fishing access. DANIEL K. RAY

In Bay City, located just above the Saginaw River's outlet into Saginaw Bay, waterfront development is adding new recreation areas and encouraging interest in the bay's rebounding sport fishery. The new Municipal Marina, with 98 boat slips and a launching ramp, is the centerpiece of the waterfront's revival. The marina also offers pedestrian access to the river atop a bulkhead and on a footpath which runs around harbor, connecting with a nearly completed six-mile Riverwalk.

The project was delayed for more than a year to resolve the cleanup of an oiled gravel road on the property, and the disposal of marina dredge spoils. Funds for the $2.4 million project came from the Michigan Department of Natural Resources, Michigan's Out-State Equity Program, the city, the county, and the federal Job Training Partnership Act.

Further up the river is a new public accessway and fishing pier. The two-mile-long riverfront footpath has interpretive displays and exercise stations. A new 700-foot-long Municipal Pier, built on the site of an old railroad bridge, provides a vista point and fishing access along the walkway. A low railing makes the fishing access suitable for wheelchairs. An additional 10 acres of existing shoreline parks along other sections of the river have been improved with new steel bulkheads and a waterfront path.

These projects' $1.6 million cost was financed by Michigan's departments of Natural Resources and Commerce, Bay County, the city, and private contributions. The Bay Area Foundation coordinated much of the private fund raising for the project, "selling" shares in the walkway light poles, benches, and other walkway features.

Private developers are adding boating facilities along the river, too. The Wheeler Landing Yacht Club, now under construction, will have 200 boat slips, a clubhouse, and swimming pool. Clean spoils dredged from the yacht basin were used to cover a nearby archaeological site, protecting the archaeological resources in the event of future development. The $4.5 million project will be privately financed. Also proposed is a 200-slip dockominium marina, Island Development Company Marina.

CONTACTS:

MIKE BRANDOW, Economic Development Specialist
City of Bay City
301 Washington Avenue
Bay City, Michigan 48708

EARL KIVISTO
Planning Department
City of Bay City
301 Washington Avenue
Bay City, Michigan 48708
(517) 894-8173

ALPENA, MICHIGAN

Alpena has enlarged its Municipal Marina on Thunder Bay. The project included dredging a boat basin and the construction of new finger piers providing 77 new boat slips. There are plans to provide bicycle and pedestrian paths in the adjacent city park.

The $1.4 million expansion was funded by the Michigan Department of Natural Resources' Waterways Fund, the Michigan Natural Resources Trust Fund, and the city. The city and the Michigan DNR's Division of Waterways and coastal zone management program shared the cost of planning the expansion.

Other recent projects include the Blair Park fishing access, a handicapped-accessible observation deck constructed atop an existing sheet piling which extends into Thunder Bay, and the North Riverfront Park boat ramps, a new two-lane launching ramp. The projects were financed by the Michigan Department of Natural Resources' Land and Water Conservation program, Michigan's coastal zone management program, and the city.

CONTACT:

GARY GRAHAM, City Engineer
City Hall
208 N. First Avenue
Alpena Michigan
(517) 354-2196

MIDLAND, ONTARIO

This Georgian Bay city has placed a moratorium on shoreline development while it prepares a new waterfront plan for its shoreline, where several grain elevators have recently closed.

Among the projects proposed is Bayport, a 26-hectare development that would feature 1,100 condominiums, a 750-slip marina (including two launching ramps, travel lift, and dry storage area for up to 200 boats), a 300-room hotel, and shopping areas scattered throughout the development. The development's condominium corporations would have first right of refusal to lease the marina slips.

Tour and party boats up to 76 metres long can be moored along the site's existing bulkhead. A footpath with vista points, and upland parks, will provide public access to the site's 670-metre shoreline. Debris from the demolition of a grain elevator now on the site may be used to create an offshore fishing reef in Midland Bay. The $200 million project will be privately financed. Another proposal seeks to redevelop the 1,100-slip Wye Heritage Marina into dockominiums, and add a hotel there.

CONTACT:

W.R. ORNER, Town Planner and Zoning Administrative Officer
City of Midland
P.O. Box 820
Midland, Ontario L4R 4P4
(705) 526-9361

COLLINGWOOD, ONTARIO

This Georgian Bay community, once known as the home of the Collingwood Shipyard, is now the site for several large resort developments. The 300-hectare Cranberry Village, now under construction, includes 750 town houses and condominiums and a 135-slip marina. Paths for walkers and bicyclists provide public access to the shoreline through the development.

Lighthouse Point, another mixed-use development on Georgian Bay, has 750 condominiums, a restaurant, and a 275-slip marina. Pedestrian and bicyclist paths which run the length of its 2.4-kilometre shoreline are for residents only. A boardwalk across a four-hectare wetland on the site allows residents to enjoy the natural setting. Both projects were privately financed.

The city is also preparing a plan for redevelopment of the now-shuttered shipyards. Among the proposals under consideration is development of a plaza and public space, as well as construction of shops and residences. The cost of the plan is being spit between the city and the Ontario Ministry of Tourism and Recreation.

CONTACT:

NANCY FARRER, Planner
City of Collingwood.
P.O. Box 157
Collingwood, Ontario L9Y 3Z5
(705) 445-1030

OWEN SOUND, ONTARIO

Owen Sound is developing its West-side Walkway, a shoreline park, on abandoned port land. The walkway is part of a phased renovation of the city's inner harbour. The first improvements to be completed were a 200-metre waterfront footpath, with lighting and landscaping. An area suitable for mooring tour boats was retained. The walkway will be extended up the east side of the harbour in the future.

The walkways are the first of a series of recreation improvements proposed for Owen Sound's inner harbour. Future projects proposed include 9,000 additional metres of waterfront footpaths, five improved vista points, a small marina, parking, two restored beaches, and landscaping.

The $801,000 committed to the project so far has come from the city and the Ontario Ministry of Municipal Affairs's PRIDE program. The Ontario Legion donated benches and decorative flags along the walkway. Cost of the total inner harbour improvement plan will be $8.7 million.

CONTACT:

STEVE HYNDMAN, City Planner
Owen Sound Planning Dept.
808 Second Avenue East
Owen Sound, Ontario N4K 2H4
(519) 376-1440

WINDSOR, ONTARIO

Windsor is now planning how to redevelopment its Riverfront Lands, about 14 hectares on the Detroit River, as a park. The site, formerly railroad right-of-way, is being acquired by the city through a land exchange between the city, the province, and the Canadian National Railway.

The railroad tracks and some buildings along the two-kilometre long right-of-way have been removed, and some minor improvements, such as landscaping, and temporary uses, such as Windsor's Freedom Festival, may proceed while a final development plan is being prepared. A 1983 plan proposed a festival plaza, marina, floral garden, museum, amphitheatre, and other recreational uses for the site. A 20-member Riverfront Lands Development Advisory Task Force, appointed by the City Council in 1987, is overseeing preparation of a new plan.

CONTACT:

STEVE LOADER, Waterfront Project Manager
City Hall
P.O. Box 1607
Windsor, Ontario N92 6S1
(519) 255-6706

SARNIA, ONTARIO

At Sarnia, improvements to boating facilities and redevelopment of port and railway lands are transforming the St. Clair River shore. Fifty new slips have been added to the Sarnia Bay Marina, with funding from Fisheries and Oceans Canada's Small Craft Harbour Branch, the Ontario Ministry of Tourism and Recreation, the cities of Sarnia-Clearwater and Chatham, and the counties of Lambton and Kent. The boat basin retains enough space for 200 more slips.

Eight acres adjoining the marina are the site for the proposed Centre for Entertainment and the Environment, an environmental interpretation and performing arts centre. It would include theatres capable of seating 750-800, restaurants, a shop, and space for exhibitions and conferences. A waterfront footpath would run along the river, linking the centre to the Sarnia Bay Marina. Planning for the project has been supported by grants from Canada's Department of Communications, the Ontario Ministry of Culture and Communications, the city, and local industries. Funding for the $24 million project would be shared by the province, the federal government, and private donations.

Also under construction on Sarnia Bay is Sarnia Civic Showplace, a 2.6-hectare park whose central feature is the MacPherson Fountain. The fully illuminated fountain will operate year-round. Future phases of the development will include a wintergarden and conservatory, with a small restaurant, several slips for transient boats, a potential tour boat berth, and landscaping. A waterfront footpath will be built along the shore. The total cost of the project is $4.7 million. The $1.92 million raised to date comes from the provincial Ministry of Community Affairs's PRIDE program, the Ministry of Tourism and Recreation's Waterfront Development Program, the city and a private contribution. The park was previously a coal yard and aggregate storage site.

Further south along the St. Clair River is the Front Street Promenade, a waterfront boardwalk with vista points, cantilevered over the adjacent Canadian National Railway right-of-way, from which strollers can watch passing ships. The $470,000 project is being funded by the provincial PRIDE program and the city.

CONTACTS:

JANE GRAHAM, Commissioner of Planning and Development
City of Sarnia-Clearwater
255 N. Christina Street
Sarnia, Ontario N7T 7N2
(519) 322-0330

A.J. DIAMOND
A.J. Diamond, Donald Schmitt and Company
Architects and Planners
2 Berkeley Street, Suite 600
Toronto, Ontario M5A 2W3
(416) 862-8800

BILL BALLARD, Commissioner
Sarnia-Lambton Economic Development Commission
155 N. Front Street
Sarnia, Ontario N7T 7V6
(519) 332-1820

GARY McDONALD, Marketing Manager
St. Clair Parkway Commission
P.O. Box 700
Corunna, Ontario N0N 1G0
(519) 862-2291

The Centre for Entertainment and the Environment, proposed for Sarnia's St. Clair River waterfront, would include theatres capable of seating up to 800 people, a shop, restaurants, and space for exhibitions and conferences. A.J. DIAMOND, DONALD SCHMITT AND COMPANY

LAKE ERIE
Including the Niagara River

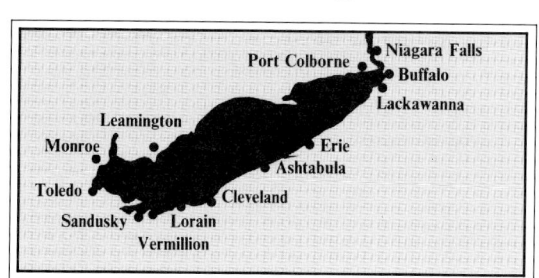

Lake Erie's waterfront revitalization has been sparked by the restoration of the Lake's water quality and fisheries, coupled with the opportunity to redevelop shorelines no longer needed for ports. The result is thousands of new boat slips, and plans for major renovations of downtown waterfronts in the region's older cities.

Projects to redevelop more than 1,000 waterfront acres have been undertaken since 1986 in the 12 Lake Erie and Niagara River cities surveyed by The Center. Almost $227 million ($C261 million) in new construction has occurred, including 13 new and expanded marinas with more than 4,400 boat slips, 739 residences, 103,000 square feet (28,700 square metres) of retail space, and 75,000 square feet (21,000 square metres) of offices. More than 165 acres (70 hectares) of waterfront parks have been added or improved, and more than 2.5 miles (four kilometres) of shoreline footpaths have been opened.

These projects are just the beginning of the revitalization planned for Lake Erie's waterfronts. Redevelopment plans are now being prepared for another 860 acres (353 hectares), adding new marinas, hotels, office towers, and parks to the lakefront. The investment in these new projects, coupled with the completion of developments now underway, is expected to be almost $1.5 billion ($C1.73 billion).

Especially noteworthy are:

- Buffalo, where the $100 million Pavilion-on-the-Lake will cap a 30-year waterfront redevelopment program.
- Cleveland, where the Flats entertainment district has brought new vitality to the banks of the Cuyahoga River.
- Sandusky, where more than 2,500 boat slips have been added in Sandusky Bay marinas.

LEAMINGTON, ONTARIO

At Leamington, waterfront development has been focused on the Municipal Marina. Its recent improvements provide 300 slips (half for transient boaters) and a launching ramp. Adding up to 250 more slips is planned, as is a new bulkhead topped by a bicycle and pedestrian path along the marina's shoreline. Additional footpaths may be constructed on a former railroad right-of-way just inland of the marina. Other planned improvements include a hotel, restaurant, visitor centre, up to 100 residences, and improvements to access roads and shoreline parks.

The total cost of redeveloping the city's waterfront is almost $20 million. About $6 million has been invested so far, including municipal funds, provincial funds (the Ministry of Municipal Affairs' PRIDE program; Ministry of Tourism and Recreation's Community Waterfront Program), and federal funds (Fisheries and Oceans Canada's Small Craft Harbours Branch).

As a condition of the provincial Ministry of the Environment's permits for the marina, the city was required to replace fish habitat damaged by the project.

CONTACT:
WILLIAM MARCK, Chief Administrative Officer
City of Leamington
38 Erie Street North
Leamington, Ontario N8H 2Z3
(519) 326-5761

PORT COLBORNE, ONTARIO

The Marina Centre Building proposed for Port Colbourne's Sugarloaf Harbours project would house shops and a restaurant. STAN SZARFLARSKI ARCHITECT LTD.

At Port Colborne, a new mixed-use development, Sugarloaf Harbours, is replacing two storm-damaged marinas at Gravelly Bay, the entrance to the Welland Canal. The project's first phase, completed in 1990, included expansion of the two marinas to a total of 980 slips, and addition of a launching ramp. The city operates 500 slips, including 100 transient slips. The remaining 480 are operated by Marlon Marina, a private company.

H.H. Knoll Lakeview Park surrounds Sugarloaf Harbours. A new one-kilometer long breakwater protects against erosion and wave attack. The second phase, a Marina Centre Building with shops and a restaurant with an outdoor patio, has just been tendered. Twenty-two condominiums have been completed in anticipation of the waterfront development. A final phase, still under consideration, may include a 70-room Marina Centre Hotel, a convention centre, and a lagoon for paddle boats in summer and ice skating in winter.

Funding for all phases of the $19 million project is expected to come from private investors, the city, the province (the Treasury Ministry's Community Economic Transformation Act program, Ministry of Tourism's Community Waterfront Development Program) and federal departments (Fisheries and Oceans Canada's Small Craft Harbours Branch and the Department of Employment and Immigration's Community Futures Program). Land for the project was assembled through long-term leases with Ports Canada, the St. Lawrence Seaway Authority, and the Port Colborne General Hospital.

CONTACT:
GRAHAM NOBLE, Economic Development Officer
City of Port Colborne
239 King Street
Port Colborne, Ontario L3K 4G8
(416) 835-2900

NIAGARA FALLS, ONTARIO

No development reported.

NIAGARA FALLS, NEW YORK

Two new developments have been proposed along the Niagara River here. The first is a new marina with 150 to 200 slips, a restaurant, water-related shops, and a municipal park. Funding for the $1-$1.5 million project is expected to be provided by a private marina developer, the city, and the New York State Environmental Quality Bond Act. The park would serve as the trailhead for the city's proposed riverfront trail.

Also proposed is a Waterfront Trail Project, a proposed park and transit system along the Niagara River. It would narrow an existing four-lane highway to two lanes, create a linear park along the River, and construct a monorail transit system within the right-of-way. The project is being proposed by the New York State Power Authority, the State's parks department, and other agencies.

CONTACTS:

RAY O'KEEFE, Economic Development Marketing Specialist
City of Niagara Falls
345 Third Street
Niagara Falls, New York 14303
(716) 286-4461

THOMAS J. DeSANTIS, Project Coordinator
City of Niagara Falls
City Hall, Room 228
745 Main Street
Niagara Falls, New York 14303
(716) 286-4470

BUFFALO, NEW YORK

Buffalo, an early leader in waterfront redevelopment, is announcing new plans to upgrade and revitalize its shoreline. Along the Niagara River shore, Buffalo is emphasizing improvement of parks. At Cornelius Creek Park, a picnic area was added along with a segment of the city's Riverwalk and a footbridge connecting the park with an adjoining boat ramp. The $750,000 project was financed with municipal general obligation bonds and a grant from New York State's Environmental Quality Bond Act of 1986.

The city's recently drafted waterfront revitalization program proposes rehabilitation of riverfront Broderick Park and creation of a new 67-acre park on the north end of Squaw Island, though toxic contamination may limit development. At nearby LaSalle Park, where the Niagara River flows out of Lake Erie, the city is planning to add transient boat slips, a fountain, and other improvements. Not far away is Lakefront Commons, a recently completed 40-unit apartment building on a former railroad right-of-way.

On the Lake Erie shore adjoining Buffalo's downtown is Waterfront Village, a 292-acre project of the Buffalo Urban Renewal Agency, which includes a marina, condominiums, and a hotel. Three residential projects have recently been constructed or are underway at Waterfront Village: the Breakwaters, a 48-unit low-rise condominium development completed in 1989; Admiral's Walk, two towers with 104 condominiums now under construction; and Portside, a 58-unit low-rise condominium.

Also under construction is Gull Landing, a mixed-use development. Its 23-unit condominium tower is already constructed. A second tower with 35 condominiums is scheduled for 1991. The third phase, still in the final design, is planned to have offices, shops, and a restaurant. A lakefront promenade will provide public access along the shore.

Final designs are also being developed for the Pavilion-on-the-Lake, the last segment of the Waterfront Village. Current plans call for 100,000 square feet of specialty and neighborhood shops and restaurants, a 12-story Buffalo World Trade Center, a 120-room Marriot Residence Inn, and a six-story parking garage. Recreation areas on the site will include pedestrian walkways, a two-acre park, and a wintergarden. The park's reflecting pool can be used as a skating rink during winter.

The Waterfront Village site was assembled by the City

Buffalo's Waterfront Village project includes several low-rise condominium projects, including the recently completed Breakwaters (foreground). CITY OF BUFFALO

and the Buffalo Urban Renewal Agency under a redevelopment plan first approved in 1963. Construction of individual projects is privately financed. Total cost of the three residential projects is more than $15.8 million. Construction costs for the first two phases of Gull Landing are $13 million. The Pavilion will cost $100 million. The Pavilion-on-the-Lake has been delayed by litigation from groups representing adjoining neighbors, who are concerned about the project's size.

New parks and recreation facilities have been added at the Buffalo River's outlet to Lake Erie, just south of Waterfront Village. Vietnam Veteran's Memorial Park, on the north side of the river, has been improved with memorial statues, 700-foot bicycle and foot paths, and a picnic area. The $400,000 project was funded with a federal Community Development Block Grant.

At the adjoining Naval and Serviceman's Park, a retired submarine, the *USS Croaker,* has been added to the park's display of historic ships, additional decks of the destroyer *USS Sullivan* have been opened to the public, and a pedestrian path has been constructed along the river. The $250,000 project was financed by revenues from admissions and concessions at the park and a loan from the City of Buffalo.

On the south side of the river, a 1,400-foot footpath has been constructed to provide access to the U.S. Coast Guard's Buffalo Lighthouse, at the mouth of the river. A picnic area on the site will be improved next. The $149,000 project is being financed by the city.

Times Beach, a former dredged spoils disposal site adjoining the lighthouse, will be managed as a wildlife refuge, with walkways and observation towers, including a handicapped-accessible observation area atop a small concession stand.

Buffalo's most recent proposal is creation of a bathing beach in the city's outer harbor on Lake Erie. To create the bathing area, a bulkhead would be constructed to enclose 50 acres of open water between the breakwater of the Niagara Frontier Transportation Agency's Small Boat Harbor and a grain elevator to the south. The enclosed lake bottom would be sealed with clay, sand would be imported to create a 15-acre beach along the shore, and water from Lake Erie would be treated and pumped into the enclosure, flooding the bathing area. A bathhouse, parking lots, and landscaping would be installed on the remaining 11 acres of the site. The plan also calls for expansion of the Small Boat Harbor itself, adding more slips and a new breakwater.

CONTACTS:
ROSANNE FRANDINA, P.E., Director of Development
City of Buffalo Department of Community Development
920 City Hall
Buffalo, New York 14202
(716) 851-5037

DAVID SENGBUSCH, Real Estate Manager
Buffalo Urban Renewal Agency
920 City Hall
Buffalo, New York 14202
(716) 851-5056

EARL DUBIN
Department of Public Works
City Hall Room 503
Buffalo, New York 14202
(716) 851-5634

COMMANDER RICHARD BECK
Buffalo Naval Servicemen's Park
1 Naval Cove
Buffalo, New York 14202
(716) 847-1773

LACKAWANNA, NEW YORK

Three hundred acres on Lackawanna's waterfront, formerly occupied by the Bethlehem Steel Company, are being redeveloped as the Gateway Trade Center, a port and industrial development. Bethlehem retains a coke processing operation on the north edge of its site, and is selectively clearing other areas. The site's port and several auxiliary buildings have been sold to the Gateway Trade Center, a subsidiary of Buffalo Crushed Stone Company which is reclaiming landfilled slag from the site.

Gateway owns two parcels: 150 acres on the north surrounding the port and 150 acres on the south. Several small industrial buildings in the northern portion are being leased by Gateway to about 25 warehouse/distribution and light manufacturing businesses which employ about 500 people. Gateway has also improved and repaved Bethlehem's port to handle general and bulk cargo, including road salt, limestone, potash, coke, and reclaimed slag. The city and Gateway are currently negotiating with Oxford Energy Company to establish a $100 million tire-to-energy plant there.

The 150 acres on the south include Bethlehem's former executive offices and Woodlawn Beach, a natural sand beach. Preliminary studies have called for rehabilitation of the beach and development of a marina.

Most of the $80 million cost of redeveloping the site will be financed by Gateway and Bethlehem. New York's legislature designated the Gateway Trade Center as an economic development zone, providing property tax abatements, investment tax credits, utility reductions, and preferential financing incentives which are intended to spur private investment and job creation. The site has also been designated a foreign trade zone by the federal government, offering a variety of opportunities to defer, reduce, or eliminate duties and excise taxes.

Lackawanna's Economic Development Zone office coordinates marketing studies, collaborates on seeking infrastructure funding for site improvements, and assists Gateway and Bethlehem in receiving planning and zoning approvals. A $250,000 grant from the Western New York Economic Development Commission was used to prepare a 1987 Waterfront Development Plan which guides the overall redevelopment of Lackawanna's waterfront.

Eight million dollars in railway track relocations and road improvements envisioned in the plan have been stalled due to the city's inability to finance them. Redevelopment is also being complicated by the U.S. Environment Protection Agency's designation of the abandoned steel plant as a Superfund site, requiring the development of a remediation plan.

CONTACT:

ROBERT DIMMIG, Director
Economic Development Zone
City of Lackawanna, City Hall, Room 309
714 Ridge Road
Lackawanna, New York 14218
(716) 827-6474

ERIE, PENNSYLVANIA

The construction of a new waterfront road, Erie Bayfront Parkway, along the shore of Presque Isle Bay is expected to spark the revitalization of Erie's bayshore. Most of the right-of-way for the three-lane, five-mile long road is former railroad lands. The road's cost is being shared by the Pennsylvania Department of Transportation and the Federal Highway Administration. The project includes measures to mitigate impacts on parks and residential neighborhoods and to deal with the possibility that abandoned hazardous waste landfills might be located along the right-of-way.

The highway is a central feature of Erie's Waterfront Comprehensive Plan. It proposes mixed-use redevelopment of 300 acres of port and industrial lands owned by the Erie-Western Pennsylvania Port Authority, the city, and a variety of private owners. The plan proposes new marinas, apartments with boat docks, parks, offices, restaurants, and shops. Erie's port facilities would be consolidated and the Port Authority's remaining lands would be redeveloped. A berth for the 19th Century sailing ship *Niagara* would be a feature of the renovated harbor. A waterfront footpath would be located along much of the bayshore.

CONTACTS:

RON DESSER, City Planner
Rm 404, Municipal Bldg.
Erie, Pennsylvania 16501
(814) 870-1234

LARRY MOROSKY, Executive Director
Erie Port Authority
17 W. Dobbins Landing
Erie, Pennsylvania 16507-1424
(814) 455-7557

ASHTABULA, OHIO

No development reported.

CLEVELAND, OHIO

On Cleveland's lively and unique waterfront, new marinas, parks, shops and restaurants contrast with port and industrial uses.

Cleveland's principal project on Lake Erie is North Coast Harbor, a 171-acre mixed-use project on land formerly used for port and railroad purposes. North Coast Harbor's first phase—an eight-acre festival park, waterfront promenade, transient marina, and infrastructure improvements—was completed in 1988.

The second phase, scheduled for completion in the mid-1990s, is planned to include four major attractions: a Great Lakes Museum of Science, Technology, and Environment, the national Rock and Roll Hall of Fame, the Cleveland Aquarium, and the *William G. Mather*, a retired ore carrier donated by the Cleveland Cliffs Company. When completed, the entire development will also include retail development and a commercial marina.

Total investment in North Coast Harbor is expected to exceed $500 million. The $11.5 million first phase was financed with municipal general obligation bonds, a state grant to build the marina, and support from North Coast Harbor, Inc. the not-for-profit development corporation which is planning and managing the project.

In the Flats entertainment district along the Cuyahoga River, the newest development is the 60-acre Nautica. Its first phase, completed in June 1987, was the renovation of an old warehouse into restaurants, nightclubs, and boating-related shops. Key attractions are Shooters Cafe, a riverside restaurant with slips where pleasure boaters can tie up, and the Nautica Stage, a 4,000-seat open-air amphitheater which hosts the Cleveland Ballet and jazz, rock, and pop artists.

Outdoor food vendors set up on the promenade along the Cuyahoga River, and the *Star of Nautica*, a Cuyahoga River excursion boat, moors there. Nautica's second phase, completed summer 1989, included the renovation of a turn-of-the-century powerhouse into offices, shops, restaurants, and nightclubs.

Future phases will include a 200-room hotel, marina, more shops and offices, and 1,500 apartments and condominiums. The marina is planned to include 600 slips and dry storage for 500 boats, as well as restaurants, marina shops and offices located in a decommissioned Coast Guard station.

The $200 million development is privately financed. The Powerhouse building is listed on the National Register of Historic Places, giving the builders and tenants a 20 percent tax credit for the $20 million in improvements to the structure.

CONTACTS:

GEORGE CANTOR
City Planning Commission
City Hall
601 Lakeside Avenue, Room 501
Cleveland, Ohio 44114
(216) 664-3807

ROBERT F. BANN II
North Coast Harbor, Inc.
1100 Chester Avenue, Suite 350
Cleveland, Ohio 44115
(216) 781-6232

LORAIN, OHIO

Marina International is Lorain's new 500-slip marina on Lake Erie. A 300-slip marina was initially planned, but before construction began the capacity was increased to 440 slips due to strong demand. The marina opened in 1989, and was expanded to 500 slips by year end. It has 12 charter fishing boat slips and a "head boat" slip for a large charter fishing/party boat holding up to 70 people. An 800-foot breakwater encloses the marina with a boat ramp adjoining it. The project is on 58-acre landfill which the city has leased from the state.

The $8.3 million project was financed with industrial revenue bonds issued by the Lorain Port Authority, which funded the boat slips, other marina facilities, and part of the breakwater, and the U.S. Army Corps of Engineers and Ohio's Department of Natural Resources, who contributed funds for the breakwater. Because Ohio law did not allow Lorain to sublease the former lake bottoms occupied by the marina, the city entered into a 32-year development agreement with the marina operator. Special approvals from the U.S. Army Corps of Engineers were required to use a federal pier for anchoring the docks and to use part of a Corps' dredged spoils disposal site for parking.

The nearby Black River Boat Launch is planned to include six launch ramps, parking, landscaping and a service building. Construction is scheduled for 1991. The launching ramp replaces an abandoned coal dock on the site. Its $1 million cost will be shared by the Lorain Port Authority and the Ohio Department of Natural Resources.

CONTACT:
RICHARD NOVAK, Executive Director
Lorain Port Authority
City Hall, Room 511
Lorain, Ohio 44052
(216) 244-2269

VERMILLION, OHIO

The resort town of Vermillion is also improving its boating facilities. The city's new Main Street Docks provide 18 berths anchored upon a bulkhead at the city's pumping station. Vermillion also rebuilt its South Street Launch Ramp, providing handicapped access, a 40-space parking lot and an access drive. The projects were financed with a grant from the Ohio Department of Natural Resources and by the Vermillion Port Authority with funds raised through user fees.

CONTACT:
WILLIAM M. SUMMER, Chairman
Port Authority
5511 Liberty Avenue
Vermillion, Ohio 44089

SANDUSKY, OHIO

Phase I of The Harbour at Sandusky includes the Harbour Marina, a Radisson Hotel, and J.J. Musky, a restaurant with 15 transient boat slips. McKAY DESIGN GROUP

At Sandusky, eight new and expanded marinas have added more than 2,500 new slips to serve pleasure boaters enjoying Sandusky Bay and the Bass Islands. There has also been more development on Cedar Point, the barrier beach separating the bay from Lake Erie.

The city has sponsored two projects. The first, three-acre Shoreline Park, includes picnic areas, fishing access piers, bayshore paths linked by foot bridges, and parking for 100 cars. Future plans call for addition of a skating rink. The site was previously an abandoned coal dock. The $770,000 project was financed from the city's general fund, a federal Community Development Block Grant, and a private donation.

The city also improved its Municipal Boat Ramp on Sandusky Bay, replacing an old hoist with a two-lane boat ramp, parking for 100 cars and trailers, a concession stand and restrooms. The cost of the half-million dollar project was shared by the city, the Ohio Department of Natural Resources' Waterways Safety Fund, and federal Sport Fish Restoration (Wallop-Breaux) Funds. Portions of the site were previously occupied by railroad right-of-way.

The largest private development on Sandusky Bay is The Harbour, a 109-acre mixed-use project. Its first phase is completed. It includes the Harbour Marina, with 230 slips, a jet ski rental, slips for charter fishing boats and party boats, shops and restaurants. An impeller system keeps the marina ice-free during winter.

The Harbour's Phase I also includes a 234-room Radisson Hotel and J.J. Musky, a waterfront restaurant with 15 transient boat slips and 198 condominiums, each with its own boat slip. Phase II will include 70 additional single-family homes, each with its own slip. The entire development is interconnected with a series of dredged canals. The $77 million project was privately financed.

Creation of the 95-acre Big Island Wetland was required by the U.S. Army Corps of Engineers to mitigate The Harbour's effects on wetlands. Big Island Wetland, created by diking off a shallow area of the bay, is managed with a system of pumps and culverts to create wetland habitat. Ohio's Sea Grant program and Ohio State University assisted in design of the mitigation program. An observation tower provides public access to the wetland for education and nature study.

Four other new marinas on Sandusky Bay include offices, shops, or boat repair yards. Battery Park has 880 slips, a restaurant and specialty shops. Five acres of the site are a public park. Its $5.2 million construction was financed by the developer, Ohio's Department of Development, and federal Community Development Block Grants and Urban Development Action Grants.

Sandusky Harbor Marina has a 700-slip marina, including two lift wells for boat launching, and dry storage for 700 boats, offices and a boat repair yard. The Murray Building will combine a 215-slip marina, now under construction, with a proposed office building. Cross Bay Storage includes a two-lane launching ramp, a jet ski rental, dry storage for 1,400 boats and a boat repair yard. The marinas at the Sandusky Yacht Club (172 new slips) and Sandusky Bay Sailing Club (30 new slips) have also been expanded. At the O and O Marina, 55 condominiums are being built and the existing 50-slip marina is being converted to dockominiums. Even the new McDonald's has slips so that the restaurant can serve boaters.

At the Cedar Point Amusement Park, new rides and attractions, shops, and a 96-room Sandcastle Suites motel have been added. Point Retreat, also on Cedar Point, has 73 condominiums, each with its own slip.

CONTACTS:

GARY BOYLE
City of Sandusky
Department of Planning and Development
222 Meigs Street
Sandusky, Ohio 44870
(419) 627-5872

ERIC METZLER
Ohio Department of Natural Resources
Division of Watercraft
Fountain Square
Columbus, Ohio 43224
(614) 265-6480

JEFF REUTTER
Ohio Sea Grant
Ohio State University
Research Center
1314 Kinnear Road
Columbus, Ohio 43212
(614) 292-8949

TOLEDO, OHIO

This port city's newest waterfront development is Commodore Island, a 29-acre mixed-use project on the Maumee River. The project, intended to bring new residents to downtown Toledo as part of ongoing revitalization efforts, is being constructed in phases. The first phase includes 83 residences (of which 15 are completed), finger piers, a yacht club, and a planned 172-slip marina. A one-acre public park is provided along the river. The construction timetable depends upon demand for the development.

The city's redevelopment agency assembled the site, installed a shoreline bulkhead, and sold the property to a private developer. The city also provided utilities, extended streets, and upgraded bridges leading to the site. Site preparation and construction of the project's $15-18 million Phase I was privately financed.

Sustaining a community consensus for the project has been difficult. Some people objected to the upscale nature of the project, complaining of its lack of housing affordable to moderate-or low-income residents, while others were concerned about restricted public access to a large portion of the riverfront. In response, the city retained the one-acre park along the river. An underground storage tank encountered during construction required clean-up before development could proceed. Additional similar problems are anticipated due to the site's previous use as a railroad freight receiving and distribution center.

CONTACT:

DON FREEMAN, Commissioner of Project Development
Department of Economic Development
One Government Center, Suite 1850
Toledo, Ohio 43604
(419) 245-1430

MONROE, MICHIGAN

A new Riverwalk will soon follow the banks of the River Raisin through downtown Monroe, at the west end of Lake Erie. The footpath and fishing access follows an existing sewer easement which runs along the river behind two blocks of the downtown's storefronts. A 50-space parking lot serves the riverwalk and the adjoining business district. The $330,000 project, now under construction, is being funded from the city's general fund.

The city's Hellenberg Field, a 13-acre park on River Raisin, has been expanded by acquiring up to 2.5 acres of adjoining land now occupied by homes. Playfields and other recreation improvements are planned for the site. The expansion was financed by Michigan's Natural Resources Trust Fund and the city. An archaeological reconnaissance of a War of 1812 battlefield on the park was required prior to installation of improvements or acquisition of additional land.

CONTACT:

ANGELA WITKOWSKI, Director
Community Economic Development
City of Monroe
120 E. First Street
Monroe, Michigan 48161
(313) 243-0700

LAKE ONTARIO

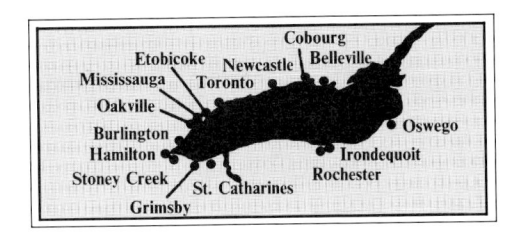

Development on Lake Ontario's coast is changing, with some Canadian communities following intensive development programs laid out years ago, and others scrambling to catch up with public concerns emphasizing more waterfront access and parks and fewer condominiums and other developments. In New York, mature manufacturing towns like Rochester and Oswego are patiently revitalizing their harbors and riverbanks to provide new marinas, recreation areas, and business sites.

Developments completed or underway on the almost 260 acres (105 hectares) of Lake Ontario shore redeveloped since 1986 include almost 2,500 condominiums, 2,000 marina slips, 72 acres (30 hectares) of parks, and more than 6.5 miles (10.5 kilometres) of waterfront walkways, with a total construction cost of more than $125 million ($C144 million). Proposed projects will add another 3,460 condominiums and apartments, 1,260 marina slips, 232,000 square feet (65,000 square metres) of shops, and 116 acres (47 hectares) of new parks with another six miles (10 kilometres) of shoreline walkways.

Especially noteworthy developments include:

- Port Dalhousie Harbour, at St. Catharines, which features a special wharf serving sport-fishing charter boats.
- Etobicoke, where gleaming condominium towers are replacing small motels along the lakefront.
- Harbourfront, the center of a major effort to reform shoreline development, led by the Royal Commission on the Future of the Toronto Waterfront.
- Metropolitan Toronto, whose regional conservation authority is building gravelly beaches from rock and concrete generated by construction projects. The new beaches provide recreation while reducing damage on eroding shorelines.
- Rochester, where the banks of the Genesee River are the focus for redevelopment of the city's downtown.

ST. CATHARINES, ONTARIO

Port Dalhousie Harbour, at the Welland Canal's outlet to Lake Ontario, is being rebuilt into a mix of marinas, shops, and parks. The Michigan Beach Marina, now under construction just outside the port's jetties, will include 350 slips, a small restaurant and marine-related shops. Two new breakwaters were built in 1990 to protect against waves and erosion.

Fisherman's Wharf, now under construction in the port, is a sportfishing marina with moorings for four sportfishing charter boats. A services building with marine-related shops, fish-cleaning and shower areas, and storage lockers for charter boat operators has been proposed. The wharf is operated by the St. Catharine's Game and Fish Association.

In Port Dalhousie's historic commercial core, two old hotels, the Murphy Building and Port Mansion, have been renovated into restaurants, night clubs, and shops. Nearby is the restored Port Dalhousie Lockup, purported to be Canada's smallest jail. Four other historic buildings in the neighborhood have been renovated with specialty shops on their ground floors and apartments on their upper levels. Hogan's Alley, an old commercial thoroughfare, is being converted to a pedestrian mall.

Other planned improvements include two new parks: Third Canal Access Park is planned for Public Works Canada land along the historic Third Canal, and Harbour Lookout Park would be on the bluff overlooking the Michigan Beach marina. A three-kilometre-long, handicapped-accessible harbour walkway would provide a footpath along the entire Port Dalhousie waterfront.

The cost of the project was estimated at $8.3 million in 1984, including both the Michigan Beach marina, funded by Fisheries and Oceans Canada's Small Craft Harbour Branch and the city, and Fisherman's Wharf, financed by the provincial Tourist Redevelopment Incentive Program and the local Game and Fish Association's Tourist Wharf Program. The walkway will require approval from Ontario Hydro, which has a control weir on the Third Canal.

CONTACT:

ROD HOLLICK, Operations Manager
Parks and Recreation Department
City of St. Catharines
P.O. Box 3012
St. Catharines, Ontario L2R 7C2
(416) 688-5600

GRIMSBY, ONTARIO

The shore of this southwestern Lake Ontario town is the site for Lakewood Gardens, a 12-lot subdivision. Revetments were installed to protect building sites from erosion hazards.

CONTACT:

WALTER BASIC, Planning Technician
Planning Department
160 Livingston Avenue
Box 159
Grimsby, Ontario L3M 4G3
(416) 945-9634

STONEY CREEK, ONTARIO

Landford Group Newport Yacht Club, under construction on 32 hectares, includes 600 condominiums, a 250-slip marina, restaurants, and shops. The restaurant is in a renovated historic estate. Agitators were installed in the marina to protect against winter ice. Public access is provided at vista points at the marina and restaurant, but the shoreline is private. The $100 million project was privately financed.

The addition of 200 boat slips has been proposed at Stoney Creek's Fifty Point Creek Conservation Area.

CONTACTS:

GARY BOYLE, Manager
Development Control
City of Stoney Creek
P.O. Box 9940
777 Highway 8
Stoney Creek, Ontario L8G 4N9
(416) 643-1261

RINO MOSTACCI, Planner
City of Stoney Creek
P.O. Box 9940
777 Highway 8
Stoney Creek, Ontario L8G 4N9
(416) 643-1261

HAMILTON, ONTARIO

Hamilton is proposing mixed-use development of a 12-hectare landfill in Hamilton Harbour. In 1991, the city will begin rehabilitation of the 1.9-hectare Pier Four Park on the site. Development of other portions of the landfill has been delayed because of toxic contamination, and because final marketing and design studies have not yet been completed.

Potential uses considered for the area include a restaurant and shops, a transient marina, and an outdoor amphitheater. Total costs of developing the project will depend strongly on final plans for the development and clean-up costs for site, now estimated at $10 million. Rehabilitation of Pier Four Park is estimated to cost $2 million.

CONTACTS:

DAVID GODLEY, Planning Department
City of Hamilton
71 Main Street
Hamilton, Ontario L8N 3T4
(416) 546-2700

KEVIN CHRISTENSON
Department of Public Works
City of Hamilton
71 Main Street
Hamilton, Ontario L8N 3T4
(416) 546-2700

BURLINGTON, ONTARIO

In Burlington, city officials are preparing plans to improve several lakefront parks. A development plan for six-hectare Shoreacres Park, purchased in 1990, will be prepared, and the master plan for Burlington Beach Waterfront Park will be reviewed, as will plans for lakefills proposed at Burloak Waterfront Park. A Great Lakes Science Center, an interpretive and research facility, is also under consideration for the city's waterfront parks.

Private development on Burlington's shoreline includes Brant's Landing, a recently completed 20-unit condominium, and Cliff House, 44 condominiums in an 11-story tower. Among the conditions of Cliff House's approval was a requirement that 25 percent of the units be affordable housing. A study of development options in the city's downtown waterfront is underway.

CONTACTS:

TIM DOBBIE, Director of Community Services
City of Burlington
P.O. Box 5013
Burlington Ontario L7R 3Z6
(416) 335-7883

RASH MOHAMMED, Planning Commissioner
Regional Municipality of Halton
P.O. Box 7000
1151 Bronte Road
Oakville, Ontario L6J 6E1
(416) 827-2151

MISSISSAUGA, ONTARIO

Mississauga's most recent waterfront improvement is the 38-hectare Lakefront Promenade Park. The park, comprised entirely of landfill, has three headlands, each with a different use.

The western headland is for passive recreation, with waterfront paths and picnic tables. The middle headland shelters a boat basin shared by two marinas: a public marina with 180 rental slips, and a 450-slip yacht club.

The eastern headland, still under construction, will provide fishing access and recreation areas. It adjoins the warm-water outfall of an Ontario Hydro thermal-electric power plant, providing year-round fishing. The park is protected from erosion and wave attack by a breakwater, revetments, and a perched beach.

The $35.6 million project was financed by the Port Credit Yacht Club, the Regional Municipality of Peel, the City of Mississauga, the Ontario Ministry of Natural Resources, and the Small Craft Harbors Branch of Fisheries and Oceans Canada. Some of the funds came from fees collected for the disposal of the construction spoils which created the park's landfill.

CONTACTS:

RICK DOYLE, Director of Parks
City of Mississauga
300 City Center Drive
Mississauga, Ontario L5B 3C1
(416) 896-5287

KEN OWEN, Manager Land Resources
Credit Valley Conservation Authority
1255 Derry Road
Mississauga, Ontario L0J 1K0
(416) 670-1615

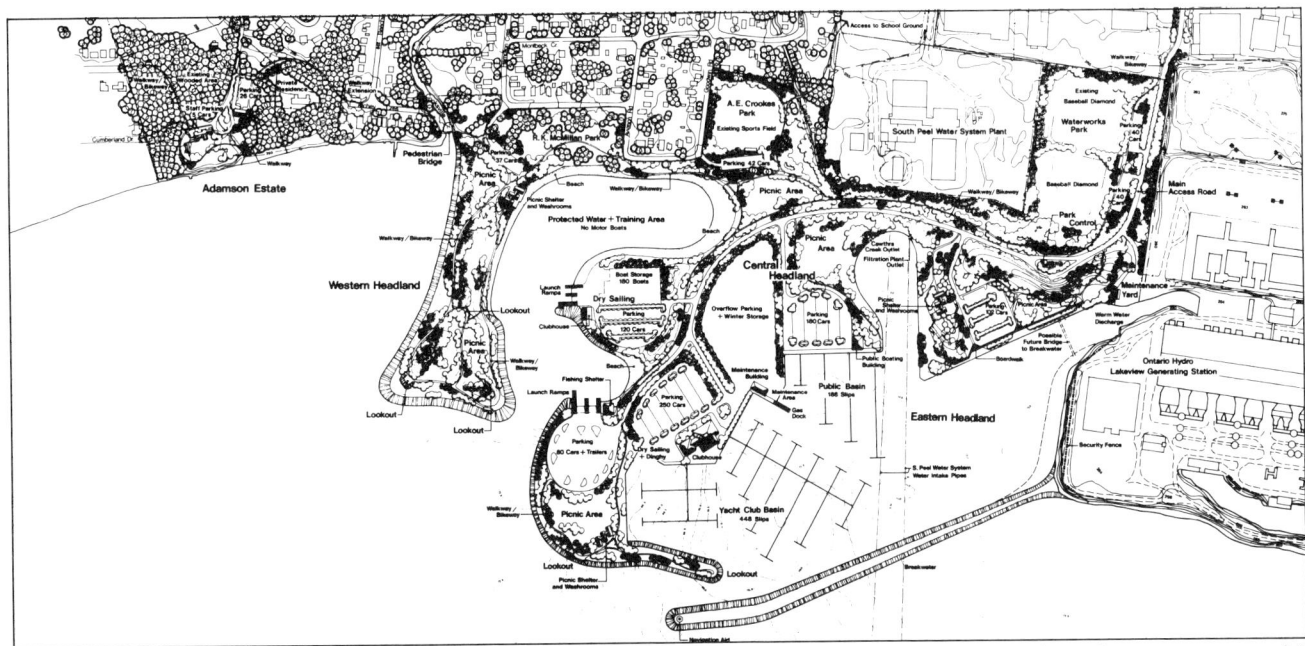

Mississauga's Lakefront Promenade Park, built entirely on landfill, features marinas and fishing access along the warmwater outfall of a power plant. VALLEY CONSERVATION AUTHORITY

OAKVILLE, ONTARIO

Bronte Harbour Waterfront Park is under construction at the mouth of Bronte Creek here. Its planned 22-hectare park will occupy both sides of the harbour at its outlet to Lake Ontario. On the park's west shore will be a bathing beach, boat launching ramp, and shoreline footpaths with a pier atop the existing harbour jetty.

On the harbour's east shore, a new 450-slip marina will be developed. A boardwalk will provide access along the marina's shore, tying into a footbridge across Bronte Creek that will link it to the park on the harbour's west shore. A waterfront promenade and park will be developed in front of the marina. The project includes nine hectares of landfill. Two new breakwaters are being built by the Small Craft Harbours Branch of Fisheries and Oceans Canada to protect against erosion and wave attack. The project is being built by the Region of Halton and will be maintained by the Town of Oakville.

At Lakeside Park, at the mouth of Sixteen Mile Creek, an artificial shingle beach has been created by constructing two armour stone hardpoints and filling the space between them with native shale from a local construction project. An armour stone retaining wall just inland of the beach creates a shoreline walkway, connecting the beach with a nearby pier and museum. A vista point on the landscaped upland portions of the park offers views of the Toronto skyline.

CONTACTS:

TED SALISBURY/BRUCE LUCYK
Town of Oakville
1225 Trafalgar Road
P.O. Box 310
Oakville, Ontario L6J 5A6
(416) 845-6601

A.J. DIAMOND
A.J. Diamond, Donald Schmidt and Company
Architects and Planners
2 Berkeley Street, Suite 600
Toronto, Ontario M5A 2W3
(416) 862-8800

ETOBICOKE, ONTARIO

Parks and condominium towers are replacing the small motels and factories which once lined this suburb's shoreline.

Colonel Sam Smith Park, now under construction, will have a 13-hectare landscaped park and a 500-slip marina. The 15-slip Harbor College Sail Center is also located here. The park has a half-kilometer long waterfront footpath. The shoreline, constructed largely of landfill, is protected from erosion by a system of hardpoints and artificial beaches. The $16 million park is a project of the Metropolitan Toronto and Region Conservation Authority and the private yacht club which operates the marina.

Three developments have been approved on Etobicoke's 20-hectare Motel Strip: a 196-unit high-rise condominium (the Andmark development) with a neighborhood shopping area, a 190-unit condominium (the Michael Ying development), and Etobicoke Harbour City—a 300-room hotel, shopping centre, and 1,037-unit condominium tower.

Parts of the Michael Ying and Etobicoke Harbour City projects are reserved for senior citizens. A waterfront promenade, built on landfill, will provide public access to the shore at the Andmark property. The intensity of development of the area and the extent of public amenities that they provide have been the source of considerable controversy.

Three more large projects are already under construction on the east end of the city's lakefront: Grand Harbour, with 20 town houses and a 440-unit condominium tower, Marina Del Rey, with 820 residences in several midrise buildings, and 46-story Palace Place, with 504 condominiums. A waterfront promenade provides public access along the lakeshore at these projects.

CONTACTS:

DAVID McKILLOP, Senior Policy Planner
City of Etobicoke
399 West Mall
Etobicoke, Ontario M9C 2Y2
(416) 394-8217

MICHAEL McKORT, Special Projects Manager
City of Etobicoke
399 West Mall
Etobicoke, Ontario M9C 2Y2
(416) 394-8228

A.J. DIAMOND
A.J. Diamond, Donald Schmidt and Company
Architects and Planners
2 Berkeley Street, Suite 600
Toronto, Ontario M5A 2W3
(416) 862-8800

LARRY FIELD
Metropolitan Toronto and Region Conservation Authority
5 Shoreham Drive
Downsview, Ontario M3N 1S4
(416) 661-6600

AJAX, ONTARIO

At Ajax, a large development proposal, the Village of Summerset Cove, has been withdrawn due to concerns about its environmental effects. The 95-hectare project would have included 2,000 single-family homes, a eight-hectare waterfront park, a marina, shops, offices, and a 200-room hotel. Among the concerns was the development's potential effects on a sensitive marsh.

CONTACT:

DAVID FORGET, Planning Technician
Ajax Planning Dept.
65 Hardwood Avenue South
Ajax, Ontario L1S 2H9
(416) 686-0360

TORONTO, ONTARIO

The federal Royal Commission on the Future of Toronto's Waterfront is developing a new focus for development of the city's shoreline, emphasizing public access and environmental protection.
HARBOURFRONT CORP.

The most exciting development on Toronto's waterfront is not a new building, but a new way of thinking about waterfront lands, advocated by the Royal Commission on the Future of Toronto's Waterfront. The Commission, created in 1988, has called for the protection of more open space and better protection of the environment along the city's shoreline, as well as improved planning for waterfronts throughout the metropolitan Toronto area. While the Commission prepares its recommendations, due out in 1991, a moratorium on waterfront development has blocked new projects on the city's shoreline.

The major ongoing work on the shore is the Toronto Harbour Commission's 466-slip Outer Harbor Public Marina, and the adjoining Tommy Thompson Park, an undeveloped 247-hectare park being created on landfill.

CONTACTS:

JOE D'ABRAMO
Waterfront and Railway Lands Section
Planning and Development Department
19 East City Hall
Toronto, Ontario M5H 2N2

SCOTT JARVEY
Metropolitan Toronto and Region Conservation Authority
5 Shoreham Drive
Downsview, Ontario M3N 1S4
(416) 661-6600

PAUL BABBS
Toronto Harbour Commission
60 Harbour Street
Toronto, Ontario M5J 1B7
(416) 863-2000

SCARBOROUGH, ONTARIO

No development reported.

NEWCASTLE, ONTARIO

In Newcastle, one new marina has just been completed, and another is being planned. The recently completed, $4.5 million Port Newcastle Marina has 250 slips. A residential subdivision is planned for 84 adjacent hectares. Both are privately financed. At Port Darlington, a newly drafted secondary plan calls for construction of a new marina, with sites for a hotel and shops and a new breakwater, east of Bowmanville Harbour. Other parts of the shore would remain undeveloped, with a footpath and new waterfront drive.

CONTACT:

CYNTHIA VAN DINTEN, Planner
Strategic Planning Branch
City of Newcastle
40 Temperance Street
Bowmanville, Ontario L1C 3A6
(416) 987-5039

COBOURG, ONTARIO

At Cobourg's 200-slip Municipal Marina, a new service building, with a fish cleaning station, showers, laundromat, and tourism services, is being added. The $500,000 building is funded by the Ontario Ministry of Tourism's Community Waterfront Development Program. Long-range plans call for the addition of up to 400 additional slips in Cobourg's outer harbour.

The city's Harbour Area Secondary Plan proposes mixed-use redevelopment of 32 hectares of industrial and railway lands adjoining the marina. The plan calls for two kilometres of new waterfront bicycle and pedestrian paths and new parks adjoining the marina, as well as retail and high-density residential uses. The plan, adopted in 1990, ends a three-year freeze on waterfront development. In response to concerns expressed by Environment Ontario, the plan specifies that development will depend on the availability of services, in particular sewage disposal capacity.

CONTACTS:

RICHARD G. STINSON, Director of Administrative Services
Town Of Cobourg
55 King Street West
Cobourg, Ontario K9A 2M2
(416) 372-4301

WAYNE G. DeVEAU, Director of Community Services
Town of Cobourg
55 King Street West
Cobourg, Ontario K9A 2M2
(416) 372-4301

BELLEVILLE, ONTARIO

Belleville's recently adopted Bayfront Planning Study proposes recreational redevelopment for 90 hectares on the Bay of Quinte and its tributary, the Moira River. The plan builds on the city's growing popularity as a sportfishing centre, proposing improvement of the 35-hectare Zwick's Island municipal park with a Bayfront Centre festival site (including a shoreline plaza and an indoor/outdoor amphitheater), a 300-slip marina, and a fishing pier.

Victoria Island municipal park would be improved with new footpaths and picnic areas, and a footbridge across the Moira River. The marina there would be redeveloped for transient boaters, with 120 slips added to the municipal marina at Meyer's Pier, where a new waterfront drive is recommended.

The municipal parks along eastern bayfront would also be improved. The privately-owned wetlands along the shoreline would be protected from development, and boardwalks and vista points would be provided near the marsh.

Along the Moira River, the plan proposes creation of a riverfront footpath in the rear lots of Front Street businesses. Fish habitats in the river would be enhanced as recommended in the Bay of Quinte Remedial Action Plan.

The plan's $12.3 million in public improvements area expected to be financed by a mix of municipal, provincial, and federal sources. Key programs expected to assist in financing the development include the Ministry of Municipal Affairs' PRIDE program, the Ministry of Natural Resources' Parks Assistance, Community Waterfront, and Community Fisheries Enhancement programs, the Ministry of Tourism and Recreation's Community Waterfront and Wintario programs, and the federal Department of Fisheries and Oceans' Small Craft Harbours Branch.

CONTACT:

STEWART MURRAY, Director of Planning
City of Belleville
169 Front Street
Belleville, Ontario K8N 2Y8
(613) 968-9543

OSWEGO, NEW YORK

Ninety new slips have been added to Wrights Landing, the municipal marina in Oswego Harbor. The $467,000 project, planned as part of the local waterfront revitalization program, was financed with revenue bonds issued by city. A new, $2.7 million Oswego Harbor Marina, with 250-300 slips, is proposed nearby. The project would be privately financed.

Two new parks, the West Side Linear Park and the East Side Esplanade, now provide public access along 1.5 miles of the Oswego River above the harbor. The parks offer fishing access and more than 2.5 miles of bicycle and pedestrian paths. Parts of the parks are built atop a newly constructed sewerage collector.

The parks, planned as part of Oswego's local waterfront revitalization program, cost $8.6 million, funded by the city, the New York Department of Environmental Conservation, a New York State Erie Barge Canal Challenge Grant, and the U.S. Department of Housing and Urban Development.

CONTACTS:

EUGENE SALOGA, Director
Community Development
McCrobie Building
41 Lake Street
Oswego, New York 13126
(315) 343-3795

ELI RAPOPORT, Assistant to the Mayor
City Hall
Oneida Street
Oswego, New York 13126
(315) 342-8140

GREGG NEAL, Waterfront Director
City of Oswego
McCrobie Building
41 Lake Street
Oswego, New York 13126
(315) 343-3795

IRONDEQUOIT, NEW YORK

Irondequoit Bay is the site of several waterfront projects. Newport Marina, now under construction, will have 187 rental slips, a restaurant and marine-related shops and services. The $2.5 million project is privately financed. Bay Tree, with 30 single-family residential lots and a dock with 30 dockominium slips, cost $1 million. Point Pleasant Estates has 16 condominiums, each with its own dock.

Improvements to New York State's Irondequoit Bay Marine Park and Monroe County's Irondequoit Bay Park West are also being considered.

CONTACTS:

KRISHAN MAGO
Town of Irondequoit
Planning and Zoning Department
1280 Fixus Avenue
Irondequoit, New York 14617
(716) 467-8840

PAUL JOHNSON, Deputy Director
Monroe County Planning Department
47 S. Fitzhugh Street, Suite 200
Rochester, New York 14614
(716) 428-5475

WATER WORKS

ROCHESTER, NEW YORK

Rochester's waterfront revitalization stretches along the Genesee River from its mouth on Lake Ontario to its intersection with the Erie Canal. Ontario Beach Park, at the Genesee River's outlet to Lake Ontario, is in the midst of a $4.5 million renovation. A new band shell and boardwalk have already been built, and the renovation of the park's historic bathhouse is planned. The improvements are financed with county general obligation bonds.

The River Street redevelopment project is planned for riverfront lands just upriver of Rochester's old port. It will include a 220-slip marina, a three-acre park, and a restaurant. Financing for the $3.5 million project has not yet been finalized.

The Urban Gorge Cultural Park interpretive center, planned for a rehabilitated water works building in the downtown Brown's Race historic district, will include a museum and restoration of an historic mill and raceway. Funding for the $2 million project comes from municipal funds and state general obligation bonds. The property was obtained through a land exchange with the local electric utility.

The 80-acre South River Corridor, between the central business district and the river's junction with the Erie Canal, is another area of riverbank redevelopment. It includes the University of Rochester campus. Four miles of foot and bicycle paths have been completed. In 1991, a new footbridge will be added to provide access across the river, linking the west bank and the university, and boating facilities for canoes, rowing skulls, and small motorcraft will be built at three locations along the river.

The first major private development, a commercial project at Brooks Avenue and Genesee Street, is expected to break ground in 1992.

Total costs of redeveloping the South River Corridor area are expected to total $100 million, including $87 million in private funds and $23 million in public funds and university expenditures. Construction costs so far have totalled $3 million. Redevelopment of the area is being undertaken through an agreement between the city, Monroe County, and the university.

CONTACTS:

WILLIAM PRICE, City Landscape Architect
City Engineer's Office
30 Church Street
City Hall, Building B 3rd Floor
Rochester, New York 14614
(716) 428-6894

THOMAS ARGUST
Rochester Parks and Recreation Department
City Hall, Rm 222-B
Rochester, New York 14614
(716) 428-6749

NANCY BURTON
Bureau of Municipal Facilities
414 Andrews Street
Rochester, New York 14604
(716) 428-7415

LARRY STID, Director
Office of Planning and Development
30 Church Street, Room 0078
Rochester, New York 14614
(716) 428-6924

Rochester's South River Corridor will include a footbridge across the river (artist's rendering). CITY OF ROCHESTER

ST. LAWRENCE RIVER

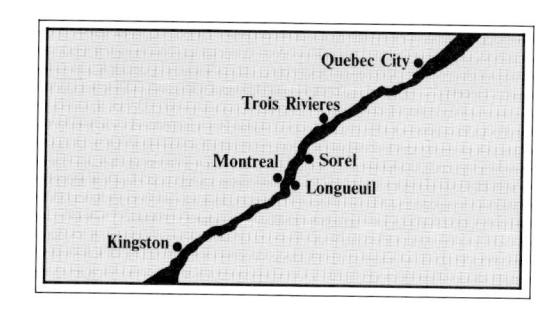

Redevelopment of the St. Lawrence River waterfront is just beginning, focusing first on a few old port areas and shorelines in metropolitan Montreal and Quebec City. Development in the past five years has included 530 marina slips, 53 acres (21 hectares) of parks, and more than five miles (eight kilometres) of waterfront paths, with a total cost of $287 million ($C332 million).

Highlights of waterfront development on the river include:

- Kingston, which bills itself as the "Freshwater Sailing Capitol of the World", with more than 1,700 boat slips, contributing more than $100 million annually to the economy.
- Montreal, where a bathing beach and wetland on Ile Notre Dame have brought nature into the middle of the city.
- Trois-Rivieres, whose Parc portuaire, an innovative waterfront terrace spanning a railroad right-of-way, combines plazas and riverside cafes with a center interpreting the pulp and paper industry, a mainstay of the region's economy.

Waterfront development focuses on bringing people back to the water. ROSS DETTMAN

KINGSTON, ONTARIO

Kingston's Flora McDonald Confederation Marina is being improved, with 176 new slips and a new outer breakwater. The enlarged boat basin created by the new breakwater has the capacity to hold up to 600 additional slips. The $4.4 million project was financed by the federal Department of Fisheries and Oceans' Small Craft Harbours Branch, the city, and the provincial Ministry of Tourism and Recreation. The Small Craft Harbors Branch also contributed to the outer breakwater's design. Transport Canada leases part of the harbour site to the city.

Kingston's system of waterfront footpaths has been extended. A 10-block long footpath has been added in a new 12-hectare park on the city's Inner Harbour on the Rideau Canal. The walkway ties into private marinas located at either end of the park. Its $200,000 construction cost was financed by the city. The second walkway is a 2.4-kilometre footpath on the Ontario Psychiatric Hospital on Lake Ontario. These paths, together with Kingston's existing waterfront parkway, provide public access along 80 percent of the city's 12-kilometre waterfront.

CONTACTS:

RUPERT DOBBIN, Director of Planning and Urban Renewal
City of Kingston
216 Ontario Street
Kingston, Ontario K7L 2Z6
(613) 546-4291

DOUGLAS FLUHRER, Commissioner of Parks
City of Kingston
216 Ontario Street
Kingston, Ontario K7L 2Z6
(613) 546-4291

LONGUEUIL, QUEBEC

Longueuil is redeveloping its riverfront, which has previously been cut off from the city by a freeway. Four kilometres of waterfront paths for pedestrians and bicycles have been built, linking two parks, parc Marie-Victorin and parc de l'île Charron, with fishing piers, a picnic area, two marinas with slips for 600 boats, a boat ramp, and a ferry to Montreal. A nature study area, Centre de la nature de la pointe Le Marigot, is located along the paths between the parks.

Three towers with 900 condominiums, a 300-room hotel, and a connection to Montreal's Metro are proposed for future construction. The $200 million development is being funded by private investors, municipal general obligation bonds, and grants from the provincial Ministry of Parks, Wildlife, and Fisheries and Office of Planning and Development. Portions of it are being built on land leased at very low cost from the St. Lawrence Seaway Authority.

CONTACT:

CLAUDE DOYON, Director of Planning
City of Longueuil
777 rue d'Auvergue
Longueuil, Quebec J4K 4Y7
(514) 646-8420

MONTREAL, QUEBEC

Montreal planners are working to turn two islands in the St. Lawrence River into Parc des Iles, an urban park, connected to the city by a ferry and subway (artist's rendering). MONTREAL URBAN COMMUNITY

The Old Port of Montreal is being made over as a 22-hectare park and recreation area. Already completed is an Imax theatre on the King Edward Quay. Other sheds at the Old Port provide sites for a flea market, festivals, and cultural exhibitions, and the wharves are used for tour boat moorings and jet ski rentals.

The western end of the historic Lachine Canal is now being rehabilitated for recreational boating. Future plans include a 10-hectare park on the Clock, Victoria, and Jacques Cartier piers, including an ice rink and warming shelters, a marina to be constructed by a private concessionaire, and improvement of the promenade and street furniture along Rue de la Commune on the edge of the Old Port. A potential Metro connection is also being considered.

Public Works Canada has pledged $100.1 million for redevelopment of the Old Port. The development is being undertaken by a crown corporation, Le Vieux-Port de Montreal.

In the river fronting the Old Port, Parc des Iles is being created by rehabilitating St. Helene's and Notre Dame Islands, where Expo '67 was held. On Ile Notre Dame, a bathing beach, bath house, sailing school, and artificial wetland have been constructed on a lagoon created on the island. The beach, wetland and surrounding area are landscaped to recreate the setting of the Laurentian lakes near Montreal. The wetland is part of the treatment system that purifies water supplied to the bathing beach. Future improvements will include restoration of the island's shoreline. The island is already the site for the city's popular winter festival.

On Ile St. Helene, Buckminster Fuller's geodesic dome, a relic of the world's fair, will be converted into an interpretive center on water and the environment, focused on the

St. Lawrence River. The island's shore will be restored, and riverside footpaths, including a vista point looking out over the city, will be added. A grassy festival site capable of holding up to 75,000 people and winter recreation facilities (including an ice rink and toboggan hill) will be provided. The ferry connecting to the mainland will be upgraded.

The projects are expected to cost $57.4 million. Financing will be shared by the city ($17.4 million) and federal ($40 million) governments.

CONTACTS:

MARK LONDON, Coordinator for Planning of the Islands
Personnel Building
St. Helene's Island
Montreal, Quebec H3C 1A9
(514) 872-1712

VICTOR LAMBERT
Le Vieux-Port de Montreal
333 Rue de la Commune
Montreal, Quebec H2Y 2E2
(514) 283-8208

SOREL, QUEBEC

Sorel has a new park, Regard sur Fleuve, with a 760-metre pedestrian path along the St. Lawrence River. It is funded by $1.65 million in municipal general obligation bonds and a $475,000 grant from the federal government. An additional $1.7 million in expenditures will be required to complete the park's improvements.

CONTACT:

GEORGE ZAKAIB, City Manager
City of Sorel
71 Rue Charlotte
Sorel, Quebec J3P 7K1
(514) 742-4583

TROIS-RIVIERES, QUEBEC

Trois-Rivieres' Parc portuaire includes a promenade along the St. Lawrence River. DAN RAY

Parc portuaire, built on an old port area, combines riverfront promenades and terraces with a museum celebrating the pulp and paper industry, one of the region's key businesses. The project's terraces include vista points and promenades, a tour boat berth, a 60-seat restaurant, meeting rooms, as well as the Pulp and Paper Interpretive Centre. Parts of the park are built atop a railroad and highway right-of-way which previously separated the river from the city centre. The $13 million project was developed and funded by Public Works Canada.

CONTACT:

PAUL LAPIERRE, Public Affairs Officer
Public Works Canada
200 W. Rene Levesque, Room 702-14
Montreal, Quebec H2Z 1X4
(514) 283-4537

QUEBEC CITY, QUEBEC

Pointe-a-Carcy, in the old port of Quebec, has had a controversial series of developments. The most recent include a festival market, restaurants, outdoor amphitheatre, and condominiums built in 1985 and 1986. In 1987, the festival market and most of restaurants were closed due to operating losses.

Recent plans for redevelopment of the site recommend demolition of the amphitheater and most of the festival market and restaurants, and restoration of the area as a park. What remains of the Grand Marche building would be renovated as a small restaurant, and the Hangar des boutiques, together with a new structure, would become a military reserve training facility. The historic Customs House would be restored as a museum. Pointe-a-Carcy's waterfront promenades and cruise ship berthing areas would be retained.

Most of these recommendations were developed by a special citizen's advisory committee. They follow years of effort to reach consensus on use of the site. The projects will be developed by the federal government, including the Department of National Defense. The naval reserve centre alone is expected to cost $30 million.

CONTACTS:

LEONCE NAUD
Secretariat a la mise en valeur du Saint-Laurent
385 Grande Allee Est
Quebec, Quebec G1R 2H8
(418) 643-7788

FRANCE GAGNON-PRATTE
Conseil des monuments et sites
C.P. 279
Haute-Ville
Quebec, Quebec G1R 4P8
(418) 694-0812

WATER WORKS

Great Lakes Waterfronts at a Glance

KEY
- ● Completed or under construction
- ◐ Some completed, some planned
- ○ Planned
- △ Funding source

	USES										BOATING					LOCAL			PROVINCE/STATE								FEDERAL				PRIVATE		
	Cultural	Hotel	Retail	Residential	Road	Industrial	Office	Parks, Beach	Nature Area	Paths/Fishing Access	Rental Marina	Dockominiums	Transient Marina	Specialty Marina	Boat Ramp	Revenue Bonds	Gen. Obligation Bonds	TIF	Other	Boating	Conservation Fund	Tourism	Urban Affairs	Economic Dev't	Infrastructure	Coastal Zone Mgmt.	Special Purpose	Other	Fisheries	Urban Affairs	Environment	Economic Dev't	Other	Builder/Developer	Foundation	Other

LAKE SUPERIOR

City	Notes
SUPERIOR, WI	Parks ●, Paths ●; △ Revenue Bonds, △ Urban Affairs, △ Builder/Developer
DULUTH, MN	Hotel ◐, Residential ●, Road ○, Industrial ●, Parks ●, Paths/Fishing ●●, Transient Marina ●; △ TIF, △ Urban Affairs, △ Economic Dev't, △ Infrastructure, △ Special Purpose, △ Builder, △ Foundation, △ Other
THUNDER BAY, ONT.	Cultural ○, Retail ○, Parks ○, Nature ○, Paths ○, Rental ○, Transient ○; △ TIF, △ Urban Affairs, △ Economic Dev't, △ Coastal Zone, △ Environment, △ Builder
SAULT STE. MARIE, ONT.	Cultural ○, Parks ●, Paths ●, Rental ●; △ Coastal Zone
SAULT STE. MARIE, MI	Retail ●, Parks ●, Nature ●, Paths ●, Specialty Marina ●, Boat Ramp ●; △ TIF, △ Boating, △ Urban Affairs, △ Builder, △ Foundation
MARQUETTE, MI	Parks ●, Paths ●, Rental ○; △ TIF, △ Urban Affairs, △ Economic Dev't, △ Other

LAKE MICHIGAN

City	
KENOSHA, WI	Retail ●, Parks ●, Paths ●●, Boat Ramp ●; △ TIF △ Other, △ Boating, △ Urban Affairs, △ Builder
RACINE, WI	Cultural ◐○, Retail ●●, Parks ◐●, Paths ●●●, Boat Ramp ●; △ Revenue, △ TIF, △ Boating △, △ Conservation, △ Builder
MILWAUKEE, WI	Cultural ●, Hotel ●, Retail ●, Residential ◐, Parks ◐●, Paths ◐●, Transient ●; △△△ local, △, △ Urban, △△△ private
SHEBOYGAN, WI	Hotel ○, Retail ●, Residential ●, Parks ●●, Paths ◐◐, Transient ●; △ TIF, △ Urban, △ Economic, △ Environment, △ Economic, △ Builder
MANITOWOC, WI	Parks ●●, Specialty ○; △ Revenue, △ Gen Ob, △ Urban, △ Environment, △ Economic
TWO RIVERS, WI	Parks ●, Transient ●, Specialty ●; △ Revenue, △ Gen Ob, △ Urban, △ Fisheries, △ Urban
STURGEON BAY, WI	Cultural ●, Parks ●●, Paths ●, Dockominiums ○, Boat Ramp ●; △ Revenue, △ Boating
GREEN BAY, WI	Parks ◐●, Paths ◐●, Boat Ramp ●; △ Revenue, △ Gen Ob, △ Urban, △ Builder, △ Foundation
MARINETTE, WI	Parks ●, Boat Ramp ●; △ Gen Ob, △ Urban, △ Economic, △ Builder, △ Foundation
MENOMINEE, MI	Parks ●; △ Revenue, △ Gen Ob, △ TIF, △ Builder, △ Foundation
ESCANABA, MI	Parks ●, Paths ●; △ Revenue, △ Gen Ob, △ TIF
CHARLEVOIX, MI	Parks ●, Nature ○, Boat Ramp ●; △ Revenue, △ Gen Ob, △ Urban
TRAVERSE CITY, MI	Parks ●, Paths ●; △ Gen Ob, △ Urban, △ Foundation, △ Other
MANISTEE, MI	Hotel ◐, Retail ●, Residential ◐, Road ○, Paths ●, Rental ●, Specialty ○; △ Rev △△ △△ Urban, △ Econ, △△ Coastal, △ Builder
MUSKEGON, MI	Retail ●, Residential ◐, Road ○, Office ●, Paths ●, Rental ◐; △ Revenue, △ Gen Ob, △ Urban, △ Economic, △ Builder
GRAND HAVEN, MI	Retail ●, Residential ●, Office ●; △ Gen Ob, △ Builder
ST. JOSEPH, MI	Retail ●, Paths ●, Transient ●, Boat Ramp ○; △ Gen Ob, △ TIF, △ Builder
NEW BUFFALO, MI	Retail ●, Office ●, Paths ●●, Dockominiums ●, Transient ○, Boat Ramp ●; △ Gen Ob, △ TIF, △ Economic, △ Builder, △ Foundation
MICHIGAN CITY, IN	Retail ●, Office ●, Transient ●; △ Tourism, △ Fisheries, △ Builder, △ Foundation
GARY, IN	Cultural ○, Hotel ○, Retail ○, Residential ○, Industrial ●, Parks ○, Paths ○○, Boat Ramp ●; △ Gen Ob, △ TIF, △ Builder, △ Foundation
EAST CHICAGO, IN	Retail ●, Paths ●, Boat Ramp ●; △ Gen Ob, △ Urban, △ Foundation
HAMMOND, IN	Retail ○, Office ○, Paths ●, Boat Ramp ●; △ Gen Ob, △ Urban, △ Environment, △ Other
CHICAGO, IL	Cultural ◐, Hotel ●, Retail ●, Residential ◐, Parks ◐●, Nature ◐, Rental ○, Transient ●●, Specialty ◐; △ Revenue, △ Gen Ob, △ TIF, △ Economic, △ Coastal, △ Builder, △ Other
HIGHLAND PARK, IL	Residential ○, Parks ○○○○; △ Revenue, △ Economic
NORTH CHICAGO, IL	Residential ○, Parks ○, Transient ○; △ Builder

LAKE HURON

City	
DETROIT, MI	Cultural ●, Retail ◐, Residential ◐, Parks ◐, Paths ●●, Transient ●, Boat Ramp ●; △ Rev △ Gen Ob, △ Fisheries, △ Urban, △ Environment, △ Builder
PORT HURON, MI	Retail ●, Residential ●, Road ●, Industrial ●, Parks ●, Transient ●, Boat Ramp ●; △ Revenue, △ Gen Ob, △ Builder
BAY CITY, MI	Parks ●, Paths ●●, Transient ○, Specialty ●●; △ Rev △ Gen Ob, △ TIF, △ Tourism, △ Economic, △ Coastal, △ Other, △ Builder, △ Foundation, △ Other
ALPENA, MI	Parks ●, Boat Ramp ●; △ Revenue, △ Gen Ob, △ TIF, △ Economic
MIDLAND, ONT.	Retail ○, Residential ○, Road ○, Parks ○, Paths ○○○, Specialty ○○; △ Builder
COLLINGWOOD, ONT.	Retail ◐, Residential ●, Road ○, Paths ◐●, Specialty ●; △ Revenue, △ Tourism, △ Builder
SARNIA, ONT.	Cultural ○, Hotel ○, Parks ●●, Paths ●, Transient ○; △ Revenue, △ Urban, △ Economic, △ Fisheries, △ Urban, △ Builder, △ Other
WINDSOR, ONT.	Cultural ○, Hotel ○, Parks ○
OWEN SOUND, ONT.	Parks ●○, Transient ●; △ Other

Great Lakes Waterfronts at a Glance

KEY
- ● Completed or under construction
- ◐ Some completed, some planned
- ○ Planned
- △ Funding source

	USES										BOATING				FUNDING LOCAL				PROVINCE/STATE									FEDERAL				PRIVATE		
	Cultural	Hotel	Retail	Residential	Road	Industrial	Office	Parks, Beach	Nature Area	Paths/Fishing Access	Rental Marina	Dockominiums	Transient Marina	Specialty Marina	Boat Ramp	Revenue Bonds	Gen. Obligation Bonds	TIF	Other	Boating	Conservation Fund	Tourism	Urban Affairs	Economic Dev't	Infrastructure	Coastal Zone Mgmt.	Special Purpose	Other	Fisheries	Urban Affairs	Environment	Economic Dev't	Other	Builder/Developer	Foundation	Other

LAKE ERIE

LEAMINGTON, ONT.		○	○	○	○		○		○	◐			●		●							△	△				△				△					
PORT COLBORNE, ONT.	○	○	●	●			●			●					●			△				△	△				△				△		△	△		
NIAGARA FALLS, NY			○		○		○		○	○								△	△					△									△			
BUFFALO, NY	●	○	○	◐		○	◐	●	◐	○		○			●		△	△	△														△			
LACKAWANNA, NY					○	●		○		○														△								△	△			
ERIE, PA	○			○	○	●		○	○			○						△	△								△									
CLEVELAND, OH	●	○	◐	○	●		◐	●		◐	○		●	◐			△			△														△	△	△
LORAIN, OH								●				●	◐	●	△			△														△				△
VERMILLION, OH								●										△	△																	
SANDUSKY, OH		●	●	◐			●	●	●	●	◐	●	●	●	△			△	△			△							△	△			△	△		
TOLEDO, OH			◐	●				○		○								△															△			
MONROE, MI								●		●								△	△																	

LAKE ONTARIO

City																																				
ST. CATHARINES, ONT.			◐	●			◐			○		●						△				△							△	△				△		
GRIMSBY, ONT.				●																														△		
STONEY CREEK, ONT.			◐	●				◐																										△		
HAMILTON, ONT.	○		○				○	○		○								△													△					
BURLINGTON, ONT.	○		●					○																										△		
OAKVILLE, ONT.							◐		◐	○			●					△						△												
MISSISSAUGA, ONT.								●		●	●							△				△										△				△
ETOBICOKE, ONT.		○	○	◐				●		◐	●							△	△												△					△
TORONTO, ONT.								●			●							△	△													△				
NEWCASTLE, ONT.			○	○	○	○			○	◐																								△		
COBOURG, ONT.			○	○			○		○	○											△															
BELLEVILLE, ONT.	○			○			○		○	○												△	△	△			△	△								
OSWEGO, NY								●		●	◐				△			△	△								△		△				△			
IRONDEQUOIT, NY			●	●				○			●	●																					△			
ROCHESTER, NY	○		○					◐		●	○					△	△												△				△	△		

ST. LAWRENCE RIVER

City																																				
KINGSTON, ONT.								●	●									△			△							△								
LONGUEUIL, QUE.		○		○				●	●	●	●			●		△		△															△		△	
MONTREAL, QUE.	◐							●	●	○	○		●					△												△		△	△			
SOREL, QUE.								●		●						△														△						
TROIS-RIVIERES, QUE.	●		●					●		●			●																	△						
QUEBEC CITY, QUE.	◐		●	●				○		○								△												△						

For Further Information

For further information on waterfront development, consider the following references and contacts:

GENERAL WATERFRONT DEVELOPMENT

Waterworks! A Survey of Great Lakes Development on U.S. and Canadian Shores, The Center for the Great Lakes, Chicago, Illinois, 1986.

Tomorrow?, Advisory Committee on the Future of Pointe-a-Carcy, Quebec, Canada, 1989.

Small Seaports, John Clark, Claudia Wilson, and Gordon Binder, The Conservation Foundation, Washington, D.C., 1979.

Urban Waterfront Lands, Ruth Fitzgerald, editor, American Society of Civil Engineers, New York, New York, 1980.

Waterfront Revitalization for Smaller Communities, Robert Goodwin, editor, Washington Sea Grant Marine Advising Services, University of Washington, 1987.

Reviving the Urban Waterfront, Andy C. Harney, editor, Partners for Livable Places, National Endowment for the Arts, and Office of Coastal Zone Management, Washington, D.C.

Improving Your Waterfront: A Practical Guide, U.S. Department of Commerce National Oceanic and Atmospheric Administration, Washington, D.C., 1980

Recreational Development of Your Community Waterfront, Video, New York Sea Grant, Oswego, New York.

Interim Report, Royal Commission on the Future of the Toronto Waterfront, Toronto, Canada, 1989.

Watershed: Interim Report, Royal Commission on the Future of the Toronto Waterfront, Toronto, Canada, 1990.

Urban Waterfronts, Ministry of Municipal Affairs—Community Planning Wing, Ontario, Canada, 1987.

Waterfront World, Waterfront Center, Washington, D.C.

Urban Waterfront Development, Douglass Wren, The Urban Land Institute, Washington, D.C., 1983.

MARINAS

Boating and Moorage in the '90s, Washington Sea Grant Marine Advisory Services, University of Washington, Seattle, Washington, 1988.

Final Report: Excursion, Cruise and Passenger Ferry Services on the Great Lakes and St. Lawrence River, Great Lakes Commission, Ann Arbor, Michigan, 1987.

Handbook for the Location, Design, Construction, Operation, and Maintenance of Boat Launching Facilities, States Organization for Boating Access, Washington, D.C., 1989.

Research Report 29: The 1987 Great Lakes Charter Sailing Industry, Minnesota Sea Grant College Program, Extension Program, University of Minnesota, Duluth, Minnesota, 1989.

National Technical Conference on Docks and Marinas, sponsored annually by the College of Engineering and the University of Wisconsin Sea Grant Institute, Madison, Wisconsin.

WORKING WATERFRONTS

Urban Ports and Harbor Management, Marc J. Hershman, editor. (Taylor and Francis, New York, 1988.)

Guidebook to the Economics of Waterfront Planning and Water Dependent Uses and *Managing the Shoreline for Water Dependent Uses: A Handbook of Legal Tools,* New England/New York Coastal Zone Task Force. (New York Dept. of State, Division of Coastal Resources and Waterfront Revitalization, 1989.)

Flats Oxbow Association, 1283 Riverbed Street, Cleveland, Ohio 44113, (216) 556-1046.

Canadian Port and Harbour Association, 60 Harbour Street, Toronto, Ontario, M5J 1B7, (416) 863-2036.

WATERFRONTS AND THE ENVIRONMENT

Making RAPs Happen: Financing and Managing Cleanups at Great Lakes Areas of Concern, The Center for the Great Lakes, Chicago, Illinois, 1991.

A Look at the Land Side: Great Lakes Shoreline Management, The Center for the Great Lakes, Chicago, Illinois, 1988.

National Wetlands Newsletter, "Wetlands of the Great Lakes," Volume 12, No. 5, Sept/Oct 1990, The Environmental Law Institute, Washington, D.C.

WINTER WATERFRONTS

Cities Designed for Winter, Jorma Manty and Norman Pressman, editors, Building Book Ltd, Helsinki, 1982.

The Winter City Book: A Guide for Survival in the Frost Belt, William Rogers.

Designs for Northern Climates, Vladimir Matus, Van Nostrand Reinhold Company, New York.

Proceedings of the Fifth International Conference on Cold Regions Engineering, C. Allen Wortley and Guenther E. Frankenstein, "Rebuilding Infrastructure for Pleasure Boating," pp. 188-201. (University of Minnesota, Oct. 1988.)

Carnaval de Quebec, Inc. 290, rue Joly, Quebec, Quebec G1L 4T8 (418) 626-3716.

Notes

1. Oscar Herrera, *Use And People's Perceptions of Waterfront Walkways: Three Case Studies in Wisconsin.* (Wisconsin Coastal Zone Management Program, Madison, 1990.)

2. Ibid.

3. The Advisory Committee on the Future of Pointe-A-Carcy, *Tomorrow* (Public Works Canada, Ottawa, 1989.)

4. The Center for the Great Lakes, *Water Works!* (Chicago, 1986.)

5. "In Search of Harbourfront", *Business Journal,* September, 1988; "Toronto on the Lake," *The Globe and Mail,* April 10, 1987; "Building Block," *The Sun,* April 11, 1987.

6. Neil De Snoo, City of Chicago Department of Planning, "Property Values and Erosion", unpublished memorandum, October, 1990.

7. Chicago Shoreline Protection Commission, *Final Report.* (City of Chicago, 1988.)

8. Cushman and Wakefield, *Business America's Real Estate Monitor.* (New York, 1989.)

9. National Council on Urban Economic Development, *Competitive Advantage: Framing a Strategy to Support High Growth Firms* (1984).

10. Richard Eisenberg, personal communication, March 1990; Richard Eisenberg and Margarite T. Smith, "The Best Places to Live in America", *Money,* (September, 1989).

11. Naud, Leonce, "Quebec: Decision-making on the Cityfront" in *Antwerp: The City and the River* (Proceedings of an International Conference, September 19-21, 1990).

12. "The Lakes Fishery: What Now?", the *Great Lakes Reporter,* November/December 1990 (The Center for the Great Lakes).

13. Douglas M. Wren, *(Urban Waterfront Development* Urban Land Institute: Washington, D.C. 1983).

14. Oscar Herrera, *Uses and People's Perceptions of Waterfront Walkways—Three Case Studies in Wisconsin: Manitowoc, Sheboygan, and Milwaukee* (Wisconsin Coastal Zone Management Program, Madison, 1990) p. 115-122.

Geographic Index

Ajax, Ontario	60
Alpena, Michigan	47
Ashtabula, Ohio	54
Bay City, Michigan	47
Belleville, Ontario	62
Benton Harbor, Michigan	40
Buffalo, New York	52
Burlington, Ontario	59
Charlevoix, Michigan	38
Chicago, Illinois	43
Cleveland, Ohio	54
Cobourg, Ontario	61
Collingwood, Ontario	48
Detroit, Michigan	46
Duluth, Minnesota	30
East Chicago, Indiana	42
Erie, Pennsylvania	54
Escanaba, Michigan	38
Etobicoke, Ontario	60
Evanston, Illinois	44
Gary, Indiana	42
Grand Haven, Michigan	40
Green Bay, Wisconsin	37
Grimsby, Ontario	58
Hamilton, Ontario	58
Hammond, Indiana	42
Highland Park, Illinois	44
Houghton, Michigan	16
Irondequoit, New York	62
Kenosha, Wisconsin	34
Kingston, Ontario	65
Lackawanna, New York	53
Leamington, Ontario	51
Longueuil, Quebec	65
Lorain, Ohio	55
Manistee, Michigan	39
Manitowoc, Wisconsin	36
Marinette, Wisconsin	38
Marquette, Michigan	32
Menominee, Michigan	38
Michigan City, Indiana	41
Midland, Ontario	47
Milwaukee, Wisconsin	13,35
Mississauga, Ontario	59
Monroe, Michigan	56
Montreal, Quebec	14,65
Muskegon, Michigan	40
New Buffalo, Michigan	41
Newcastle, Ontario	61
Niagara Falls, New York	51
Niagara Falls, Ontario	51
North Chicago, Illinois	44
Oakville, Ontario	60
Oswego, New York	62
Owen Sound, Ontario	48
Port Colborne, Ontario	51
Port Huron, Michigan	46
Quebec City, Quebec	17,66
Racine, Wisconsin	15,34
Rochester, New York	63
Sandusky, Ohio	55
Sarnia, Ontario	49
Sault Ste. Marie, Michigan	32
Sault Ste. Marie, Ontario	32
Scarborough, Ontario	61
Sheboygan, Wisconsin	36
Sorel, Quebec	66
St. Catharines, Ontario	58
St. Joseph, Michigan	40
Stoney Creek, Ontario	58
Sturgeon Bay, Wisconsin	37
Superior, Wisconsin	31
Thunder Bay, Ontario	31
Toledo, Ohio	19,56
Toronto, Ontario	61
Traverse City, Michigan	38
Trois-Rivieres, Quebec	66
Two Rivers, Wisconsin	37
Vermillion, Ohio	55
Waukegan, Illinois	44
Wilmette, Illinois	44
Windsor, Ontario	48

Ordering Waterfront Development Fact Sheets

Detailed Fact Sheets describing the waterfront development projects profiled in this report are available from The Center for the Great Lakes. The Fact Sheets are a product of The Center's Great Lakes Information Service, which provides information and reference and referral services on a wide variety of Great Lakes/St. Lawrence River issues to media, government agencies, corporations, and citizen groups.

Each waterfront development Fact Sheet includes a detailed description of a specific project, profiling the development's features, its location, and prior uses of the development site. A detailed breakdown on the project's finances is included, outlining the development's cost and listing the principal sources of funding. Planning, zoning, and permitting approvals required of each project are also covered. The Fact Sheets summarize key issues affecting major projects, and briefly highlight innovative approaches used to carry out the development. Contacts for further information, including city officials, developers, and consultants, are provided.

The Fact Sheets are intended to help waterfront builders, city officials, planners, engineers, and architects confronting the challenge of revitalizing Great Lakes/St. Lawrence River waterfronts. They can be used to learn how builders in other communities in the region have successfully redeveloped their shores and overcome barriers that are common throughout the basin.

The matrix on pages 67 and 68 is intended as an aid to those seeking information on specific projects or about common types of waterfront projects and challenges. The matrix identifies each city profiled, the status of its projects, the uses developed or proposed, and the principal sources of funding. Use the matrix to identifying the other communities in the region which share waterfront development features of most concern to you. Fact Sheets on projects in these cities can then be ordered by the name of the development.

To order, use the order form below. Indicate the Fact Sheets desired, complete the personal information on the order form, and return it to the Great Lakes Information Service at The Center's Chicago or Toronto offices. Please remit payment when ordering and make checks payable to The Center for the Great Lakes. For bulk orders or to order Fact Sheets on all projects in a given city, please call the Information Service at (312) 263-0785 or (416) 921-0776.

FACT SHEET ORDER FORM

WATERFRONT DEVELOPMENT FACT SHEETS: List the projects for which you want Fact Sheets. SINGLE COPIES are free. There is a 25 cent charge for multiple copies, and a postage/handling charge of $2.00 for orders of at least 24 Fact Sheets and $3.00 for 50 or more. Please include payments when ordering.

TO ORDER ALL FACT SHEETS FOR A PARTICULAR CITY, or to place large orders, call The Great Lakes Information Service, at (312) 263-0785 or (416) 921-0776.

_____   _____
_____   _____
_____   _____
_____   _____

FOR A FACT SHEET ABOUT THE CENTER FOR THE GREAT LAKES, check here: _____

PLEASE COMPLETE THE FOLLOWING INFORMATION:

Name, Title: _____   Amount enclosed: _____

Organization: _____

Address: _____

City: _____ Province/State: _____ Postal/ZIP code: _____

Check the categories that best describe your affiliation with Great Lakes issues:

- ____ Academic
- ____ Citizens' Group
- ____ Corporate
- ____ Elected Government
- ____ Foundation
- ____ Interested Individual
- ____ News Media
- ____ Non-academic Research
- ____ Non-elected Government
- ____ Professional Association/Union
- ____ Student
- ____ Other: _____

MAIL THIS FORM TO:

The Center for the Great Lakes
Information Service
35 E. Wacker Drive, Suite 1870
Chicago, IL 60601

The Centre for the Great Lakes
Information Service
320½ Bloor St. West, Suite 301
Toronto, ON M5S 1W5

Date Due

BRODART, INC.　　Cat. No. 23 233　　Printed in U S A

MICHIGAN CHRISTIAN COLLEGE LIBRARY
ROCHESTER, MICH.